CRITERIA FOR
ENERGY PRICING POLICY

CRITERIA FOR
ENERGY PRICING POLICY

A collection of papers commissioned for the Energy Pricing Policy Workshop
organized under the Regional Energy Development Programme (RAS/84/001),
Bangkok, 8-11 May 1984.

Sponsored by the United Nations Development Programme (UNDP), the Economic and
Social Commission for Asia and the Pacific (ESCAP), the International Labour
Organization (ILO), the Commission for European Communities (CEC), the East–West
Center (EWC), and the International Development Research Centre (IDRC)

Edited by
Corazón Morales Siddayao

Published by
Graham & Trotman

The opinions expressed herein are those of the authors and do not necessarily reflect the views of the United Nations.

First published in 1985 by

Graham & Trotman Ltd
Sterling House
66 Wilton Road
London SW1V 1DE
UK

Graham & Trotman Inc.
13 Park Avenue
Gaithersburg
MD 20877
USA

British Library Cataloguing in Publication Data

Energy Policy Pricing Workshop *(1984: Bangkok)*
Criteria for energy pricing policy: a collection of papers commissioned
for the Energy Pricing Policy Workshop organised under the Regional
Energy Development Programme (RAS/84/001), Bangkok, 8-11 May
1984.
1. Power resources — Prices — Government policy — Developing
countries
I. Title   II. Siddayao, Corazón Morales   333.79'17   HD9502.D/

ISBN 978-94-011-9812-7   ISBN 978-94-011-9810-3 (eBook)
DOI 10.1007/978-94-011-9810-3

Typeset in Great Britain by Electronic Village Ltd, Richmond

CONTENTS

Preface

Chapter 1 ENERGY PRICING POLICY FRAMEWORK AND
EXPERIENCE IN DEVELOPING COUNTRIES
Mohan Munasinghe

Chapter 2 SOCIO-ECONOMIC GOALS IN ENERGY PRICING POLICY:
A FRAMEWORK FOR ANALYSIS
Manmohan S. Kumar

APPENDIXES

PREFACE

The main part of this volume is composed of papers commissioned for the Energy Pricing Policy Workshop held at Bangkok from 8 to 11 May 1984, co-ordinated by the United Nations Economic and Social Commission for Asia and the Pacific (ESCAP) and the Resource Systems Institute of the East–West Center. The Workshop, which involved high-level policy planners from several Asian developing countries, was financially sponsored by several organizations: the ESCAP Regional Energy Development Programme funded by the United Nations Development Programme; the European Economic Community; the International Labour Organisation; and the International Development Research Centre. Publication of these papers has been supported by funds from the European Economic Community and the United States Agency for International Development. Preparation of the manuscripts for publication was undertaken at the East–West Center, through contribution in kind at the professional, editorial, and support staff levels.

The Workshop developed out of discussions at the Eighth Session of the ESCAP Committee on Natural Resources, 27 October to 2 November 1981, where questions on resource pricing in general (as related to energy resources) and sectoral pricing policies for end-users were discussed in the context of demand management. It was recognized that, although much work had been done in the area, policy makers could seldom obtain analytically supported, yet realistic, guidance concerning energy pricing questions. The emphasis of the Workshop was, therefore, on policy decision-making, and the application of theory to policy formulation in the energy pricing area. Detailed follow-up studies to be done in some of the countries represented were planned, as the report in Appendix I indicates.

The papers in this collection are concise revisions of the original documents presented at the Workshop. In some instances, revisions have been substantive. The collection is intended to serve policy makers interested in understanding the role of pricing policy. Economic policy, while ideally applying economic theory, involves the choices of policy makers to intervene directly in the workings of the economy to improve its performance, as well as the choice not to make that direct intervention to allow the different economic factors to respond to economic forces freely. It also involves how that intervention — if chosen as an approach — should take place, or the choice of the instrument variables to be employed. Energy policy is a part of economic policy insofar as it affects the production, supply, and use of energy; hence, it cannot be designed independently of economic policy, although some aspects of it may be related to other non-economic policies.

Thus, appropriate energy policy has the following characteristics:
(a) It must be integrated with a country's other goals, e.g. its balance of payments position, energy security to assure the achievement of economic development targets, the achievement of macro-economic goals (investment, income, and employment), as well as goals related to science and technology, the environment, and socio-economic development.
(b) In light of the first characteristic, energy policy must be evaluated in the context of efficiency in the allocation of resources to increase overall welfare as well as equity in sharing that welfare increase.
(c) Incorporating efficiency and equity objectives in attempting to achieve the overall objectives of energy policy cannot be evaluated without linking such objectives to the individual factors that would respond to such policy.

This collection is not intended to be a cookbook, for while all authors are in agreement with the above starting points, the complexity of the subject is reflected in some divergence of opinions in some aspects of pricing policy. Of necessity, some repetition of ideas may occur where certain issues are critical to the arguments presented.

This volume caps the first stage of a joint effort by the organizations involved. It would, however, not have been possible without the outstanding co-operation of everyone involved. Thanks are also due the support staff involved in this effort, both at ESCAP and at the East–West Center. Editing and preparation of the manuscripts for publication required the careful attention of and timely co-ordination by Dorothy Izumi, Helen Takeuchi, Jennifer Cramer, and Sonya Ho. Our many thanks to all who helped expedite preparation of this volume.

Corazón M. Siddayao
Honolulu
May 1985

Chapter 1

ENERGY PRICING POLICY FRAMEWORK AND EXPERIENCE IN DEVELOPING COUNTRIES

Mohan Munasinghe

INTRODUCTION

Today's societies require increasing amounts of energy for domestic, industrial, commercial, agricultural, and transport uses. These energy needs are met by the commercial energy sources including the short-term, depletable fossil fuel supplies — petroleum, coal, and natural gas — as well as the long term, renewable sources — hydroelectric, biomass, solar, geothermal, wind, and tidal power (Munasinghe and Schramm, 1983).

This paper sets out a consistent generalized framework for energy pricing in developing countries. The methodology seeks to maintain a compromise between analytical rigour and practicality. Because energy pricing is only one aspect of demand management and overall energy planning for national development, it is useful to first examine the role of pricing policy within this wider perspective.

PRICING POLICY AND INTEGRATED NATIONAL ENERGY PLANNING

Because of the many interactions and nonmarket forces that shape and affect the energy sectors of every economy, decision makers in an increasing number of countries have realized that energy sector investment planning, pricing, and management should be carried out on an integrated basis, e.g., within a national planning framework which helps analyse energy policy options ranging from a short-run supply-demand management to a long-run natural energy strategy (Munasinghe, 1983). However, in practice, most policies are still carried out on an *ad hoc* and, at best, regional, partial, or subsectoral basis. Thus, typically, electricity and oil subsector planning have traditionally been carried out independently of each other as well as of other energy subsectors. Environmental planning has focused on the pollution effects of energy systems but has given little attention to the resulting consequences in terms of alternative choices of energy resources and the overall costs of these choices to the economy. As long as energy was relatively cheap, such partial approaches and the resulting economic losses were acceptable, but lately, with rising energy costs (especially of oil), drastic changes in relative fuel prices, and increasing substitution possibilities, the advantages of more integrated energy policies have been evident.

Co-ordinated energy planning and pricing require detailed analyses of the interrelationships between the various economic sectors and their potential energy requirements on the one hand (Munasinghe, 1980c), and of the capabilities and advantages and disadvantages of the various energy sectors such as electric power, petroleum, natural gas, coal, and traditional fuels (e.g., firewood, crop residues, and dung) to satisfy these requirements on the other. Nonconventional sources, whenever they turn out to present viable alternatives, must also be fitted into this framework. The discussion applies both to the industrial and the developing world. In the former, the complex and intricate relationships between the various economic sectors, as well as the prevalence of private market decisions on both the energy demand and the supply sides make analysis and forecasting of policy consequences a difficult task. In the latter, substantial levels of market distortions, shortages of foreign exchange and human and financial resources for development, larger numbers of poor households whose basic needs somehow have to be met, greater reliance on traditional fuels, and relative paucity of energy, as well as other considerations add to the complicated problems faced by energy planners everywhere.

Demand management and pricing policy

Supply-and-demand management makes it easier for the energy policy-maker to forecast and achieve energy supply-demand balances, thus preventing major economic disruptions and consequent reductions in national welfare. Supply management includes identification and optimal exploitation of all energy resources, investment planning, transformation, and refining and distribution of energy. Demand management includes all means of influencing the magnitudes and patterns of energy consumption. As discussed later in this chapter, the so-called "hard tools" of demand management such as physical controls and rationing, mandatory regulations relating to the pattern of energy production and use, and technological options such as energy-saving retrofits are most effective in the shorter term. The "soft" tools of demand management such as pricing, taxation, financial incentives and subsidies, and education and propaganda are more useful in the medium and long run.

In order to understand the important role of pricing, we first clarify the scope of integrated national energy planning and demand management by examining the hierarchical framework depicted in Figure 1.1. At the highest and most aggregate level, it must be clearly recognized that the energy sector is a part of the whole economy. Therefore, energy planning requires analysis of the links between the energy sector and the rest of the economy. Such links include the impact on the economy of policies concerning prices, taxes, and availability in relation to national objectives and the input requirements of the energy sector such as capital, labour, raw material, and environmental resources such as clean air, water, or space, as well as energy outputs such as electricity, petroleum products, and woodfuel.

While some of these relationships are at the macro-level — such as foreign exchange requirements for energy imports, or investment capital requirements for the energy sector — others are more directly linked with and limited to specific activity levels. For example, price-related policies affecting the transport sector, such as subsidies to public transport, construction or nonconstruction of superhighways or airports, the levels of licence fees for vehicles or excise taxes on diesel versus gasoline vehicles, tax credits for energy conservation, pollution control legislation, or specific end-use planning policies, may have as profound an impact on energy demands as more overall broad-based energy pricing, allocation, or supply management policies.

The second level of integrated national energy planning treats the energy sector as a separate entity composed of subsectors, such as electricity and petroleum products. This permits detailed analysis of each sector with special emphasis on interactions among the different energy subsectors, substitution possibilities, and the resolution of any resulting policy conflicts such as competition between natural gas, bunker oil or coal for electricity production, diesel or gasoline use in transport, kerosene and electricity for lighting, or woodfuel and kerosene for cooking.

The third and most disaggregate level pertains to planning within each of the energy subsectors. Thus, for example, the electricity subsector must determine its own demand forecast, long-term investment programmes and price; the petroleum subsector, its supply sources, refinery outputs, distribution networks, and likely demands for oil products; the woodfuel subsector, its consumption projections and detailed plans for rotation or reforestation, and harvesting of timber.

In practice, the three levels of integrated national energy planning merge and overlap considerably. For example, a class of demand management issues that affects both macro and micro aspects of energy planning are those related to energy substitutions or energy conservation. Within certain limits many energy resources are substitutes for each other, although price, convenience in use, and overall systems cost may vary widely. Hence, appropriate supply and pricing policies may bring about significant shifts in energy demand for specific energy resources, at least in the long run.

Similarly, individual actions or deliberate policies aimed at bringing about energy conservation — e.g., reductions in energy usage relative to levels that would prevail in their absence — may significantly affect energy consumption. Such conservation may simply be achieved at the expense of some loss in personal comfort or convenience (like reducing thermostat settings, driving within mandated speed limits, or switching off lights in unoccupied rooms). Other means may consist of the substitution of energy by capital or labour, the replacement of pilot lights by electronic switches, the reduction in the curb weight of automobiles, recirculation of process heat in industrial plants through better engineering or lighter materials, or the installation of insulating materials in buildings.

Policy tools and constraints

To achieve the desired objectives of energy planning and energy demand management, the policy tools available to a government for optimal supply-demand planning and management include: (1) pricing; (2) physical controls; (3) technical methods (including research and development); (4) direct investments or investment-influencing tax policies; and (5) education and promotion. Pricing is the most effective tool of demand management, especially in the medium and long run, and the remainder of this chapter analyses this aspect in detail. The scope of the other policy instruments are discussed elsewhere (Munasinghe, 1980b). Since these tools are inter-related, their use should be closely co-ordinated for maximum effect.

In the context of developing countries, we generally face additional constraints on energy policies, especially pricing. There may be severe market distortions due to taxes, import duties, subsidies, or externalities which cause market (or financial) prices to diverge substantially from the true economic opportunity costs, or shadow prices. Therefore, on the grounds of economic efficiency alone we may have to make (second-best) shadow pricing adjustments. However, these again may have to be modified in anticipation of energy user reactions that will be based on market prices rather than underlying economic cost considerations. Furthermore, there often are severe income disparities and social considerations which call for subsidized energy prices or rationing to meet the basic energy needs of poor consumers. Finally, there are usually many additional considerations that affect policy decisions, such as considerations of future investment requirements, financial viability and autonomy of the energy sector, regional development needs, as well as socio-political, legal, and other constraints.

SCOPE AND OBJECTIVES OF PRICING POLICY

Energy pricing is a very important tool for demand management, especially in the long run. As discussed below, the pricing and investment decisions should be closely related. However, energy supply systems — e.g., electricity generation, transmission, and distribution; oil and gas wells and pipelines; coal mines; and forests—usually require large capital investment with long lead times and lifetimes. Therefore, once the investment decision is made, usually on the basis of the conventional least-cost method of meeting demand by subsector, with due regard for interfuel substitution possibilities, there is a lock-in effect with respect to supply. Thus, prices should be related to the long-run planning horizon. On the demand side also, energy conversion devices (e.g., motor cars, gas stoves, electric appliances, and machines) are expensive relative to average income levels and have relatively long lifetimes, thus limiting consumers' ability to respond in the short run to changes in relative fuel prices.

The objectives of energy pricing are closely related to the goals of energy planning, but they are more specific. First, the economic growth objective requires that pricing policy should promote economically efficient allocation of resources, both within the energy sector and between it and the rest of the economy. In general terms, this implies that future energy use would be at optimal levels, with the price (or the consumer's willingness to pay) for the marginal unit of energy used reflecting the incremental resource cost of supply to the national economy. Relative fuel prices should also influence the pattern of consumption in the direction of the optimal or least-cost mix of energy sources required to meet future demand. Distortions and constraints in the economy necessitate the use of shadow prices and economic second-best adjustments, as described in the next section.

Second, the social objective recognizes every citizen's basic right to be supplied with certain minimum energy needs. Given the existence of significant numbers of poor consumers and also wide disparities of income, this implies subsidized prices, at least for low-income consumers.

Third, the government would be concerned with financial objectives relating to the viability and autonomy of the energy sector. This would usually be effected by pricing policies that permit institutions (typically, government owned) in the different energy subsectors to earn a fair rate of return on assets and to self-finance an acceptable portion of the investments required to develop future energy resources.

Fourth, energy conservation is also an objective of pricing policy. While prevention of unnecessary waste is an important goal, other reasons often underlie the desire to conserve certain fuels. These include the desire for greater independence from foreign sources (e.g., oil imports) and the necessity of reducing the consumption of woodfuel because of deforestation and erosion problems.

Fifth, we recognize a number of additional objectives, such as the need for price stability, to prevent shocks to consumers from large price fluctuations, and the need for simplicity in energy pricing structures, to avoid confusing the public and to simplify metering and billing.

Finally, there are other specific objectives, such as promoting regional development (e.g., rural electrification) or specific sectors (e.g., export-oriented industries), and other socio-political, legal, and environmental constraints.

In summary, therefore, price is most effective as a long-run policy tool. From the viewpoint of economic efficiency, the price indicates to suppliers the consumers' willingness to pay and the use value of energy; to the consumers, it signals the present and future opportunity costs of supply that draws on various energy sources.

Role of government in pricing policy

We conclude this section with a brief review of the pervasive role that most governments play in the pricing of commercial energy resources and the relative neglect of issues relating to traditional forms of energy. Governments exercise direct influence, usually through the ownership of energy sources or price controls. Indirect influences occur through such means as taxes, import duties, subsidies, market quotas, taxes on energy-using equipment, and government-guided investments in energy resources.

In practically all developing countries, the electric utility is government owned. In oil and gas production, refining, and distribution, as well as in coal mining, both public and private organizations often operate side by side. However, irrespective of the form of ownership, all governments exercise some form of wholesale or retail price control, usually at several levels, including during production, during refining, after transport or transmission, and so on. Income and excise taxes are also levied from both public and private energy sector companies.

Generally, certain fuels in specific uses tend to be subsidized, although leakages and abuses of subsidies by nontargeted consumer groups also occur. Thus, kerosene for lighting and cooking, rural electricity for lighting and agricultural pumping, and diesel fuel for transportation commonly qualify for subsidies. Cross-subsidies exist between different fuels, user groups, and geographic regions; therefore high-priced gasoline may finance the subsidy on kerosene, industrial electricity users may subsidize household consumers, and a uniform national pricing policy usually implies subsidization of energy users in remote areas by those living in urban centres. The principal problem associated with subsidies is that the energy producer may not be able to raise sufficient revenues to finance investment to meet expanding demand, or even to maintain existing facilities, and thus shortages eventually result. Furthermore, cross-subsidies give consumers the wrong price signals, with consequent misallocation of investments.

Import and export duties, excise taxes, and sales taxes are levied, often by several levels of government, from federal to municipal, at various stages in the production, processing, distribution, and retailing chain. In many developing countries, the combined levies are several hundred percent of the original product price for some items, and negative or close to zero for others. Several less obvious methods, such as property taxes, water rights and user charges, and franchise fees are also used to influence energy use. Energy prices are also affected by the wide range of royalty charges, profit-sharing schemes, and exploration agreements that are made for the development of oil and gas resources between governments and multinational companies.

Other policy instruments are often used to reinforce pricing policies, such as quotas on imported or scarce forms of energy, coupled with high prices. Conservation regulations may affect depletion rates for oil and gas, while the availability of hydropower from some multipurpose dams may be subordinate to the use of water for irrigation or river navigation. Many special policies involving tax holidays and concessions, import subsidies, export bonuses, government loans or grants, high taxes on large automobiles, etc., are also used to affect energy use.

The traditional fuels subsector has been relatively neglected because transactions involving these forms of energy are usually of a noncommercial nature. However, there is growing acceptance of the co-ordinated use of indirect methods such as displacement of fuelwood used in cooking by subsidizing kerosene and liquefied petroleum gas (LPG), increasing the supply of fuelwood by reafforestation programmes and effective distribution of charcoal, enforcing stiffer penalties for illegal felling of trees, and proper watershed management.

ECONOMIC FRAMEWORK AND BASIC PRICING MODEL

Because the objectives mentioned above are often not mutually consistent, a realistic integrated energy pricing structure must be flexible enough to permit trade-offs among them. To allow this flexibility, the formulation of energy pricing policy must be carried out in two stages. In the first stage, a set of prices that strictly meets the economic efficiency objective is determined, based on a consistent and rigorous framework. The second stage consists of adjusting these efficient prices (established in the first stage) to meet all the other objectives. The latter procedure is more *ad hoc*, with the extent of the adjustments being determined by the relative importance attached to the different objectives. In the rest of this section, we discuss the importance of shadow pricing and develop the economic framework that permits the efficient pricing of energy. The second stage adjustments due to noneconomic factors are discussed in the next section.

Shadow pricing theory has been developed mainly for use in the cost-benefit analysis of projects (Mishan, 1976). However, since investment decisions in the energy sector are closely related to the pricing of energy outputs, for consistency the same shadow pricing framework should be used in both instances. Shadow prices are used instead of market prices (or private financial costs) to represent the true economic opportunity costs of resources (see Chapter 6 by Siddayao).

In the idealized world of perfect competition, the interaction of atomistic profit-maximizing producers and atomistic utility-maximizing consumers yields market prices that reflect the correct economic opportunity costs, and scarce resources including energy will be efficiently allocated. However, in the real world, distortions may result from monopoly practices, external economies and diseconomies (which are not internalized in the private market), interventions in the market process through taxes, import duties, and subsidies, etc., and these distortions cause market prices for goods and services to diverge substantially from their shadow prices or true economic opportunity costs. Therefore, shadow prices must be used in investment and output pricing decisions to ensure the economically efficient use of resources. Moreover, if there are large income disparities, we will see later that even these "efficient" shadow prices must be further adjusted, especially to achieve socially equitable energy pricing policies for serving poor households.

It is important to realize that lack of data, time, and manpower resources, particularly in the least developed countries context, will generally preclude the analysis of a full economy-wide model when energy-related decisions are made (Little and Mirrlees, 1974; Squire and van der Tak, 1975; Munasinghe, 1979). Instead, the partial approach shown in Figure 1.1 may be used, whereby linkages and resource flows between the energy sector and the rest of the economy, as well as interactions among different energy subsectors, are selectively identified and analysed, using appropriate shadow prices such as the opportunity cost of capital, shadow wage rate, and marginal opportunity cost for different fuels. In practice, surprisingly valuable results may be obtained from relatively simple models and assumptions.

To clarify the basic concepts involved in optimal energy pricing, we first analyse a relatively simple model. Next, the effects of more complex features are examined, including short-run versus long-run dynamic considerations, capital indivisibilities, joint output cost allocation, quality of supply, and price feedback effects on demand. The process of establishing the efficient economic price in a given energy subsector may be conveniently analysed in two steps (Munasinghe, 1980b). First, the marginal opportunity cost (MOC) or shadow price of supply must be determined. Second, this value has to be further adjusted to compensate for demand-side effects arising from distortions in the price of other goods, including other energy substitutes. From a practical viewpoint, an optimal pricing procedure that begins with MOC is easier to implement, because supply costs are generally well defined (from technological-economic considerations), whereas data on the demand curve are relatively poor.

Suppose that the marginal opportunity cost of supply in a given energy subsector is the curve MOC(Q) shown in Figure 1.2. For a typical non-traded item like electricity, MOC that is generally upward sloping is calculated by first shadow pricing the inputs to the power sector and then estimating both the level and structure of marginal supply costs (MSC) based on a long-run system expansion programme (Munasinghe, 1981). For tradable items like crude oil and for fuels that are substitutes for tradables at the margin, the international or border prices of the tradables (i.e., c.i.f. price of imports or f.o.b. price of exports, with adjustments for internal transport and handling costs) are appropriate indicators of MOC. For most developing countries, such import or export MOC curves will generally be flat or perfectly elastic. Other fuels such as coal and natural gas could be treated either way, depending on whether they are tradables or nontraded. The MOC of nonrenewable, nontraded energy sources will generally include a "user cost" or economic rent component, in addition to the marginal costs of production. The economic values of traditional fuels are the most difficult to determine because in many cases there is no established market. However, as discussed later, they may be valued indirectly on the basis of the savings they allow on alternative fuels such as kerosene, the opportunity costs of labour for gathering firewood, and/or the external costs of deforestation and erosion.

Thus, for a nontraded form of energy, MOC is the opportunity cost of inputs used to produce it plus a user cost where relevant, while for a tradable fuel or a substitute, MOC represents the marginal foreign exchange cost of imports or the marginal export earnings foregone. In each case, MOC measures the shadow-priced economic value of alternative output foregone because of increased consumption of a given form of energy. After identifying the correct supply curve, we next examine demand-side effects, especially second-best corrections that capture interactions between different energy subsectors. This second step is just as important as the first one, and therefore it will be examined in some detail.

In Figure 1.2 the market-priced demand curve for the form of energy under consideration is given by the curve PD(Q), which is the consumers' willingness to pay. Consider a small increment of consumption dQ at the market price level P. The traditional optimal pricing approach attempts to compare the incremental benefit of consumption due to dQ, that is, the area between the demand curve and X-axis, with the corresponding supply cost, that is, the area between the supply curve and X-axis. However, since MOC is shadow priced, PD must also be transformed into a shadow-priced curve to make the comparison valid. This is done by taking the increment of expenditure P.dQ and asking "what is the shadow-priced marginal cost of resources used up elsewhere in the economy if the amount P.dQ (in market prices) is devoted to alternative consumption (and/or investment)?"

Suppose that the shadow cost of this alternative pattern of expenditure is b(P.dQ), where b is called a conversion factor. Then the transformed PD curve, which represents the shadow costs of alternative consumption foregone, is given by b.PD(Q); in Figure 1.2, it is assumed that b < 1. Thus, at the price P, incremental benefits EGJL exceed incremental costs EFKL. The optimal consumption level is Q_{opt}, where the MOC and b.PD curves cross, or equivalently where a new pseudo-supply curve MOC/b and the market demand curve PD intersect. The optimal or efficient selling price to be charged to consumers (because they react only along the market demand curve PD, rather than the shadow-priced curve b.PD) will be Pe=MOC/b at the actual market clearing point B. At this level of consumption, the shadow costs and benefits of marginal consumption are equal, that is, MOC=b.PD. Since b depends on user-specific consumption patterns, different values of the efficient price Pe may be derived for various consumer categories, all based on the same value of MOC. We clarify the foregoing by considering several specific practical examples.

First, suppose that all the expenditure (P.dQ) is used to purchase a substitute fuel, that is, assume complete substitution. Then the conversion factor b is the relative distortion or ratio of the shadow price to market price of this other fuel. Therefore P = MOC/b represents a specific second-best adjustment to the MOC of the first fuel, to compensate for the distortion in the price of the substitute fuel. For example, MOC could represent the long-run marginal cost of rural electricity (for lighting), and the substitute fuel could be imported kerosene. Suppose that the (subsidized) domestic market price of kerosene is not at one half its import (border) price for socio-political reasons. Then b=2, and the efficient selling price of electricity P = MOC/2 (ignoring differences in the quality of the two fuels, and capital cost of conversion equipment such as light bulbs, kerosene lamps, and partial substitution effects; a more refined analysis of substitution possibilities would have to incorporate these additional considerations). It would be misleading, however, to then attempt to justify the subsidized kerosene price on the basis of comparison with the newly calculated low price of electricity. Such circular reasoning is far more likely to occur when pricing policies in different energy subsectors are unco-ordinated, rather than in an integrated energy pricing framework. We note that all these energy sector subsidies must be carefully targeted to avoid leakages and abuses, as discussed in the next section.

Next, consider a less specific case in which the amount (P.dQ) is used to buy an average basket of goods. If the consumer is residential, b would be the ratio of the shadow price to the market price of the household's market basket (here, b is also called the consumption conversion factor). The most general case would be when the consumer was unspecified, or detailed information on consumer categories was unavailable, so that b would be the ratio of the official exchange rate (OER) to the shadow exchange rate (SER), which is also called the standard conversion factor (SCF). This represents a global second-best correction for the divergence between market and shadow prices averaged throughout the economy. For example, suppose the border price of imported diesel is 4 pesos per litre (i.e., US$0.20 per litre, converted at the OER of 20 pesos per US dollar). Let the appropriate SER that reflects the average level of import duties and export subsidies be 25 pesos per US dollar. Therefore, SCF=OER/SER=0.8, and the appropriate strictly efficient selling price of diesel is P = 4/0.8=5 pesos per litre.

Extensions of the basic model

The analysis so far has been static. However, in many instances the situation with regard to the availability of a given energy source, interfuel substitution possibilities, and so on, tends to vary over time, thus leading to disequilibrium in certain fuel markets and divergence of the short-run price from the long-run optimal price. This aspect is illustrated below by means of an example that shows how the optimal depletion rate and time path for MOC of a domestic nonrenewable resource will be affected by varying demand conditions, especially tradability, extent of reserves, and substitution possibilities.

Suppose that the present-day marginal supply cost or MSC (including extraction costs, additional transport and environmental costs, etc., where appropriate) of a domestic energy source such as coal lies below the thermal equivalency price of an internationally traded fuel (e.g., petroleum or high-quality coal), as indicated by points A and B in Figure 1.3. The international energy price that acts as the benchmark is assumed to rise steadily in real terms, along the path BE. Let us first examine two polar extremes based on simple, intuitively appealing arguments.

First, if the reserves are practically infinite and the use of this fuel at the margin will not affect exports or substitution for imports of traded fuels, then the MOC of the domestic energy source in the long run would continue to be based on the marginal supply cost, that is, along the path AC, which is upward sloping to allow for increases in real factor costs or extraction costs. On the other hand, suppose there is a ready export market for the indigenous resource, or substitution possibilities with respect to imported fuels. In this case the marginal use of this resource will reduce export earnings or increase the import bill for the international fuels in the short run, because the reserves are small or output capacity is limited. Then, the marginal opportunity cost would tend to follow the path AD and rise quickly toward parity with the international energy price.

The actual situation is likely to fall between these two extremes, thus yielding alternative price paths such as AFE, or AGHE. Here, the initial use of the resource has no marginal impact on exports or import substitution, but there is gradual depletion of finite domestic reserves over time, and eventual transition to higher-priced fuels in the future. For a given volume of reserves, the rate of depletion of the domestic energy source will be greater, and the time to depletion will be shorter if its price is maintained low (i.e., on the path AGHE) for as long as possible rather than when the price rises steadily (i.e., along path AFE). The macro-economic consequences of the path AGHE are also more undesirable because of the sudden price increase at the point of transition, when the domestic resource is exhausted. In practice, the price path may well be determined by noneconomic factors. For example, the price of newly discovered gas or coal may have to be kept low for some years to capture the domestic market and displace the use of imported liquid fuels (which continue to be subsidized for political reasons). In general, the desire to keep energy prices low as long as possible must be balanced against the need to avoid a large price shock in the future.

The preceding discussion is more useful for all importing or energy-deficit developing countries. In the case of major oil exporters, the ability to influence the world market price and to determine the rate of resource depletion provides much greater flexibility. The huge foreign exchange surpluses and limited capacity to absorb investment imply decreased attractiveness of marginal export earnings coupled with the need to conserve oil resources (Samii, 1979). There is also greater ability to subsidize domestic oil consumption to meet basic needs and to accelerate economic development by increasing investment and expanding nonoil gross domestic product.

More rigorous dynamic models which maximize the net economic benefits of energy consumption over a long period, have been developed to determine the optimal price path and depletion rate; however, these models depend on factors such as the social discount rate, the size of reserves, the growth of demand, and the cost and time lag needed to develop a backstop technology (which could replace the international energy price as the upper bound on price). Uncertainties in future supply and demand — such as the possibility of discovering new energy resources or technologies — add to the complexities of dynamic analysis. The classical argument developed by Hotelling (1938) indicates that the rate of increase in the optimal rent (or difference between price and marginal extraction cost) for the resource should equal the rate of return on capital, r (in our case this would be the social discount rate). This implies that the optimal path of MOC would be IJE in Figure 1.3, defined at any time t by

$$MOC(t) = MSC(t)JL/(1r)T\text{-}t$$

where JL is the rent at the time of depletion T. Thus, MOC consists of the current marginal costs of extraction, transport, environmental degradation, and so on (MSC), plus the appropriately discounted "user cost" or foregone surplus benefits of future consumption (JL). As T approaches infinity, IJ would tend toward AC, which is the infinite reserve case, while as T falls to zero, IJ would approximate AD more closely, corresponding to the case of very small reserves and rapid transition to the expensive fuel.

We now consider another type of dynamic effect due to the growth of demand from year 0 to year 1, which leads to an outward shift in the market demand curve from D0 to D1 as shown in Figure 1.4. Assuming that the correct market clearing price P0 was prevailing in year 0, excess demand equal to GK will occur in year 1. Ideally, the supply should be increased to D1 and the new optimum market clearing price established at P1. However, the available information concerning the demand curve D may be incomplete, making it difficult to locate the point L.

Fortunately, the technical-economic relationships underlying the production function or known international prices usually permit the marginal opportunity cost curve to be determined more accurately. Therefore, as a first step, the supply may be increased to an intermediate level Q', at the price p'. Observation of the excess demand MN indicates that both the supply and, if necessary, also the marginal cost price should be further increased. Conversely, if we overshoot L and end up in a situation of excess supply, then it may be necessary to wait until the growth of demand catches up with the oversupply. In this iterative manner, it is possible to move along the MOC curve toward the optimum market clearing point. As we approach it, note that the optimum is also shifting with demand growth, and therefore we may never hit this moving target. However, the basic guideline of pegging the price to the marginal opportunity cost of supply and expanding output until the market clears is still valid.

Next, we examine the practical complications raised by price feedback effects. Typically, a long-range demand forecast is made assuming some given future evolution of prices, a least-cost investment programme is determined to meet this demand, and optimal prices are computed on the basis of the latter. However, if the estimated optimal price that is to be imposed on consumers is significantly different from the original assumption regarding the evolution of prices, then the first-round price estimates must be fed back into the model to revise the demand forecast and repeat the calculation.

In theory, this iterative procedure could be repeated until future demand, prices, and MOC estimates become mutually self-consistent. In practice, uncertainties in price elasticities of demand and other data may dictate a more pragmatic approach in which the MOC would be used to devise prices after only one iteration. The behaviour of demand is then observed over some time period and the first-round prices are revised to move closer to the optimum, which may itself have shifted as described earlier.

When MOC is based on marginal production costs, the effect of capital indivisibilities or lumpiness of investments causes difficulties in many energy subsectors. Thus, owing to economies of scale, investments for electric power systems, gas production and transport, oil refining, coal mining, reforestation, and so on tend to be large and long lived. As shown in Figure 1.5, suppose that in year 0 the maximum supply capacity is $\bar{Q}$, while the optimal price and output combination (P, Q) prevails, corresponding to demand curve Do and the short-run marginal supply cost curve (SRMSC) (e.g., variable, operating, and maintenance costs).

As demand grows from Do to D1 over time and the limit of existing capacity is reached, the price must be increased to P to clear the market — that is, "price rationing" occurs. When the demand curve has shifted to D2 and the price is P2, capacity is increased to $\bar{Q}$. However, as soon as the capacity increment is completed and becomes a sunk cost, price should fall to the old trend of SRMSC — for example, P3 is the optimum price corresponding to demand D3. Generally, the large price fluctuations during this process will be disruptive and unacceptable to consumers. This practical problem may be avoided by adopting a long-run marginal cost (LRMC) approach, which provides the required price stability while retaining the basic principle of matching willingness to pay and incremental supply costs. Essentially, the future capital costs of a single project or an investment programme are distributed over the stream of output expected during the lifetime of this plant. This average investment cost per unit of incremental output is added to variable costs (SRMSC) to yield LRMC, as shown in Figure 1.5.

Another method of allocating capacity costs, known as peak load pricing, is particularly relevant for electricity and also natural gas. The basic peak load pricing model shown in Figure 1.6 has two demand curves; for example, Dpk could represent the peak demand during the X daylight and the evening hours of the day when electric loads are large, while Dop would indicate the off-peak demand during the remaining (24-X) hours when loads are light. The marginal cost curve is simplified assuming a single type of plant with the fuel, operating, and maintenance costs given by the constant a, and the incremental cost of capacity given by the constant b. The static diagram has been drawn to indicate that the pressure on capacity arises due to peak demand Dpk, while the off-peak demand Dop does not infringe on the capacity $\overline{Q}$. The optimal pricing rule now has two parts corresponding to two distinct rating periods (i.e., differentiated by the time of day):

peak period price   Ppk = a + b
off-peak period price   Pop = a

Alternatively, we seek a schedule of prices which maximizes net benefits (NB), equal to total revenue plus consumer surplus minus operating costs, subject to the constraint that output (demand) does not exceed the capacity limit: i.e., Maximize $NB = {}_0\int^Q P(Q)dQ - C(Q)$ subject to $Q < \overline{Q}$; or equivalently;

optimize the Lagrangian $L = NB - m(Q - \overline{Q})$. (1)

where Q = output, P = price, C = operating cost, $\overline{Q}$ = capacity limit, and m is the Lagrange multiplier.

The first-order condition for optimizing equation (1) with respect to Q implies that:

$P = a + m$ . . . . (2)

where $a = \partial C / \partial Q$ is the marginal energy or operating cost.

That is, when the capacity constraint is not binding (i.e., $Q < \overline{Q}$ and m = O), price should equal the marginal energy (or operating) cost of producing the utility's output. During periods of peak demand when the capacity constraint is effective (i.e., $Q = \overline{Q}$ and m = b), the optimal price increases to P = (ab). In this case b equals the cost of relaxing the capacity constraint by one unit and has the conventional interpretation of unit capacity cost. In general, measures of capacity cost should be forward-looking and reflect the per unit cost of meeting a sustained increment in peak period demand. Consumers' willingness to pay prices equal to (a + b) signals that capacity expansion is economically justified. The logic of this result is that peak period users, who are the cause of capacity additions, should bear full responsibility for the capacity costs as well as fuel, operating, and maintenance costs, while off-peak consumers pay only the latter costs. Peak-load pricing can also be applied in different seasons of the year.

Related problems of allocating joint costs arise in other energy subsectors as well — an example is the allocation of capacity costs of natural gas, or of refinery costs among different petroleum products. The former may be treated like the electricity case. For oil products, the light refinery costs that are tradable, such as kerosene, gasoline, and diesel, have benchmark international prices. However, other items like heavy residual oils may have to be treated like nontradables. Furthermore, associated gas that may be flared at the refinery is often assumed to have a low MOC, although subsequent storage and handling for use as LPG will add to the costs. A more complicated approach would be to use a programming model of a refinery to solve the dual problem as a means of determining shadow prices of distillates.

The interrelated issues of supply and demand uncertainty, safety margins, and shortage costs also raise complications (Munasinghe, 1980a). We first illustrate this issue using electricity as an example, and then generalize the results for the other subsectors. Thus, the least-cost system expansion plan to meet an electricity demand forecast is generally determined assuming some (arbitrary) target level of system reliability — e.g., loss-of-load probability (LOLP), reserve margin. Therefore, marginal costs depend on the target reliability level, when in fact economic theory suggests that reliability should also be treated as a variable to be optimized, and both price and capacity (or equivalently, reliability) levels should be optimized simultaneously. The optimal price is the marginal cost price as described earlier, while the optional reliability level is achieved when the marginal cost of capacity additions (to improve the reserve margin) are equal to the expected value of economic cost savings to consumers due to electricity supply shortages averted by those capacity increments. These considerations lead to a more generalized approach to system expansion planning, as shown below (Munasinghe and Gellerson, 1979).

Consider a simple expression of the net benefits (NB) of electricity consumption, which are to be maximized:

$$NB(D,R) = TB(D) - SC(D,R) - OC(D,R)$$

where TB = total benefits of consumption if there were no outages; SC = supply costs (i.e., system costs); OC = outage costs (i.e., costs to consumers of supply shortages); D = demand; and R = reliability.

In the traditional approach to system planning (i.e., least-cost system expansion planning), both D and R are exogenously fixed, and therefore NB is maximized when SC is minimized. However, if R is treated as a variable,

$$d(NB)/dR = -\partial(SC+OC)/\partial R + [\partial(TB\text{-}SC\text{-}OC)/\partial D] \cdot (\partial D/\partial R) = 0$$

is a necessary first-order maximization condition.
Assuming $\partial D/\partial R = O$, we have:

$$\partial(SC)/\partial R = -\partial(OC)/\partial R$$

Therefore, as described earlier, reliability should be increased by adding to capacity until the above condition is satisfied. An alternative way of expressing this result is that since TB is independent of R, NB is maximized when total costs, $TC=(SC+OC)$, are minimized. The above criterion effectively subsumes the traditional system planning rule of minimizing only the system costs (Munasinghe, 1980d).

We note that this approach may be generalized for application in other energy subsectors. Thus, while sophisticated measures of reliability like LOLP do not exist outside the power subsector, the concept of minimizing total costs to society is still relevant. For example, in oil and gas investment planning, the costs of shortages due to gasoline queues, lack of furnace oil, or gas for domestic and industrial use may be traded off against the supply costs of increased storage capacity and greater delivery capability incurred by augmenting surface transport or pipeline systems. Clearly, these additional considerations would modify the marginal costs of energy supply and thus affect optimal pricing policies.

Finally, externalities, especially environmental considerations, have to be included as far as possible in the determination of efficient energy prices. For example, if the building of a new hydroelectric dam results in the flooding of land that has recreational or agricultural value, or if urban transportation growth leads to congestion and air pollution, these costs should be reflected in MOC (Seneca and Taussig, 1979). While such externality costs may, in certain cases, be quite difficult to quantify, they may already be included (at least partially) on the supply side, in terms of measures taken to avoid environmental degradation (e.g., the cost of pollution control equipment at an oil refinery or coal-burning electricity plant, or the cost of landscaping strip-mined land).

Estimation of environmental costs is most problematic in the case of noncommercial or traditional energy sources such as woodfuel, where marginal opportunity costs could be based (when appropriate) on the externality costs of deforestation, erosion, loss of watershed, and so on. Other measures of the economic value of traditional fuel would include the opportunity cost of labour required to collect woodfuel, or the cost savings from displaced substitute fuels such as kerosene and LPG.

ADJUSTMENTS TO EFFICIENT PRICES TO MEET OTHER OBJECTIVES

Once efficient energy prices have been determined, the second stage of pricing must be carried out to meet social, financial, political, and other constraints.

We note that efficient energy prices deviate from the prices calculated on the basis of financial costs because shadow prices are used instead of market prices. This is done to correct for distortions in the economy. Therefore, the constraints that force further departures from efficient prices (in the second stage of the pricing procedure) may also be considered as distortions that impose their own shadow values on the calculation (Munasinghe and Warford, 1982).

Subsidized prices and lifeline rates

Socio-political or equity arguments are often advanced in favour of subsidized prices or "lifeline" rates for energy, especially where the costs of energy consumption are high relative to the incomes of poor households. Economic reasoning based on externality effects may also be used to support subsidies, for example, cheap kerosene to reduce excessive firewood use and prevent deforestation, erosion, and so on. To prevent leakages and abuse of such subsidies, energy suppliers must act as discriminating monopolists. Targeting specific consumer classes (for example, poor households) and limiting the cheap price only to a minimum block of consumption are easiest to achieve, in practice, for metered forms of energy like gas or electricity. Other means of discrimination, such as rationing and licensing, may also be required (Munasinghe, 1980b). All these complex and interrelated issues require detailed analysis.

The concept of a subsidized "social" block or "lifeline" rate for low-income consumers has another important rationale, based on the income redistribution argument. We clarify this point with the aid of Figure 1.7, which shows the respective demand curves for energy AB and GH of low (I_L) and average (I_A) income domestic users, the social tariff Ps over the minimum consumption block O to Qmin and the efficient price level Pe. All tariff levels are in domestic market prices. If the actual price $P = Pe$, the average household will be consuming at the "optimal" level Q_A, but the poor household will not be able to afford the service.

If increased benefits accruing to the poor have a high social value, then, although in nominal domestic prices the point A lies below Pe, the consumer surplus portion ABF multiplied by an appropriate social weight w could be greater than the shadow price of supply. The adoption of the block tariff shown in Figure 1.7, consisting of the lifeline rate Ps, followed by the full tariff Pe, helps capture the consumer surplus of the poor user but does not affect the optimum consumption pattern of the average consumer — for example, a minimum ration of cheap electricity or kerosene to poor households.

In practice, the magnitude Qmin has to be carefully determined to avoid subsidizing relatively well-off consumers; it should be based on acceptable criteria for identifying "low-income" groups and reasonable estimates of their minimum consumption levels (e.g., sufficient to supply basic energy requirements for the household). The level of Ps relative to the efficient price may be determined on the basis of the poor consumer's income level relative to some critical income level. On the basis of a simplified model, (Munasinghe, 1980b) it may be shown that: $Ps = MOC$ [I_L/(critical income or poverty line)]. The financial requirements of the energy sector would also be considered in determining Ps and Qmin. This approach may be reinforced by an appropriate supply policy (e.g., subsidized house connections for electricity and special supply points for kerosene).

Financial viability

The financial constraints most often encountered relate to meeting the revenue requirements of the sector and are often embodied in criteria such as some target financial rate of return on assets or an acceptable rate of contribution towards the future investment programme. In principle, for state-owned energy suppliers, the simplest solution would be to set the price at the efficient level and to rely on government to subsidize losses or tax surpluses exceeding sector financial needs. In practice, some measure of financial autonomy and self-sufficiency is an important goal for the sector. Because of the premium that is placed on public funds, a pricing policy that results in failure to achieve minimum financial targets for continued operation of the sector would rarely be acceptable. The converse and more typical case, where efficient pricing would result in financial surpluses well in excess of traditional revenue targets, might be politically unpopular, especially for an electric utility. Therefore, in either case, changes in revenues have to be achieved by adjusting the efficient prices.

It is intuitively clear that discriminating between the various consumer categories, so that the greatest divergence from the marginal opportunity cost-based price occurs for the consumer group with the lowest price elasticity of demand, and vice versa, will result in the smallest deviations from the "optimal" levels of consumption consistent with a strict efficiency pricing regime (Munasinghe and Schramm, 1983). In many countries the necessary data for the analysis of demand by consumer categories is rarely available, so rule-of-thumb methods of determining the appropriate tariff structure have to be adopted. However, if the energy subsector exhibits increasing costs (i.e., if marginal costs are greater than average costs), the fiscal implications should be exploited to the full. Thus, for example, electric power tariffs (especially in a developing country) constitute a practical means of raising public revenues in a manner that is generally consistent with the economic efficiency objective, at least for the bulk of the consumers who are not subsidized; at the same time they help supply basic energy needs to low-income groups. Similar arguments may be made in the petroleum subsector, where high prices for gasoline, based on efficiency, externality, and conservation arguments, may be used to cross-subsidize the "poor man's" fuel — kerosene or diesel used for transportation (Munasinghe, 1980b). However, a number of undesirable side-effects may follow, such as the practice of mixing gasoline with kerosene and the substitution of diesel for gasoline. The income distribution effects may also be perverse, with the relatively wealthy diverting cheap kerosene or diesel for use in vehicles or in industry.

Other considerations

There are several additional economic, political, and social considerations that may be adequate justification for departing from a strict efficient pricing policy. The decision to provide commercial energy like kerosene or electricity in a remote rural area (which often also entails subsidies because the beneficiaries are not able to pay the full price based on high unit costs) could be made on completely noneconomic grounds (e.g., for general socio-political reasons such as maintaining a viable regional industrial or agricultural base, stemming rural to urban migration, or alleviating local political discontent). Similarly, uniform nationwide energy prices are a political necessity in many countries, although this policy may, for example, imply subsidization of consumers in remote rural areas (where energy transport costs are high) by energy users in urban centres. However, the full economic benefits of such a course of action may be much greater than the apparent efficiency costs that arise from any divergence between actual and efficient price levels. Again this possibility is likely to be much more significant in a developing country than in a developed one, not only because of the high cost of energy relative to incomes in the former, but also because the available administrative or fiscal machinery to redistribute incomes (or to achieve regional or industrial development objectives by other means) is frequently ineffective.

The conservation objective (to reduce dependence on imported energy, improve the trade balance, and so on) usually runs counter to subsidy arguments (see also Chapter 3, by Newbery). Therefore, it may be necessary to restrict cheap energy to productive economic sectors that need to be strengthened, while in the case of the basic energy needs of households, the energy price could be sharply increased for consumption beyond appropriate minimum levels. In other cases, conservation and subsidized energy prices may be consistent. For example, cheap kerosene might be required, especially in rural areas, to reduce excessive woodfuel consumption and thus prevent deforestation and erosion.

It is particularly difficult to raise prices to anywhere near the efficient levels where low incomes and a tradition of subsidized energy have increased consumer resistance. In practice, price changes have to be gradual, in view of the costs that may be imposed on those who have already incurred expenditures on energy using equipment and made other decisions, while expecting little or no change in traditional energy pricing policies. At the same time, a steady price rise will prepare consumers for high future energy prices. The efficiency costs of a gradual price increase can be seen as an implicit but not easily quantifiable shadow value placed on the social benefits that result from this policy.

Finally, owing to the practical difficulties of metering, price discrimination, and billing, and the need to avoid confusing consumers, the pricing structure may have to be simplified. Thus, the number of customer categories, rating periods, consumption blocks, and so on, will have to be limited. Electricity and gas offer the greatest possibilities for structuring. The degree of sophistication of metering depends, among other things, on the net benefits of metering and on problems of installation and maintenance. In general, various forms of peak electricity pricing (i.e., using maximum demand or time-of-day metering) would be particularly applicable to large- and medium-sized commercial consumers as well as high-voltage industrial consumers. However, for very poor consumers receiving a subsidized rate for electricity, a simple current limiting device may suffice, because the cost of even simple kWh metering may exceed the net benefits (which equal the savings in supply costs due to reduced consumption, less the decrease in consumption benefits). For electricity or gas, different charges for various consumption blocks may be effectively applied with conventional metering. However, for liquid fuels like kerosene, subsidized or discriminatory pricing would usually require schemes involving rationing and coupons and could lead to leakage and abuses.

RECENT EXPERIENCE IN ASIA AND THE PACIFIC

In this section we examine representative examples of recent energy prices prevailing in developing countries in Asia and the Pacific. Particular emphasis is given to electricity and petroleum product prices because the complexities of price structuring are greatest in these two energy subsectors.

Power tariffs used by utilities, which have borrowed from the Asian Development Bank, have been converted into US cents at the official exchange rate and are summarized in Table 1.1. Bearing in mind the overvaluation of local currency and other conversion problems, caution is required in interpreting the data.

Table 1.1 shows that in 1982 the average tariff ranged from 2.51¢/kilowatt hour (kWh) in Afghanistan to 25.40¢/kWh in the Solomon Islands. Five utilities charged less than 5¢/kWh, while for five others the average rate was above 12¢/kWh. Several factors such as the generation plant mix, government policies on subsidizing fuel prices, rural electrification and industrial promotion, and the financial objectives of the utilities are important in explaining the variation in the level and structure of tariffs in different Asian and Pacific developing countries.

The table shows that 21 utilities out of 26 had demand and energy charges; 16 utilities had a lifeline rate; 9 employed fuel cost adjustment clauses; and only 7 used time-of-day pricing — generally, for large industrial and commercial consumers. The use of time-of-use tariffs will probably increase in the future as more of the utilities undertake tariff studies based on marginal costs and implement the results. In recent years such studies have been undertaken in a number of Asian and Pacific countries, including Bangladesh, Burma, Indonesia, Republic of Korea, Pakistan, Papua New Guinea, the Philippines, Sri Lanka, and Thailand. As a result of these studies, Indonesia, Republic of Korea, Pakistan, and Sri Lanka have already changed their tariffs to reflect marginal costs, and both Bangladesh and Burma are in the process of doing so. Modifications in the electricity price structures of other countries have also been influenced by such studies.

Relative price changes of petroleum products since 1973 are given in Table 1.2. These fuels are used predominantly in the transport, industrial, commercial, and household sectors. The comparable price indices for Saudi Arabian marker (light crude) ex-Ras Tanura were: 1973 = 100, 1977 = 459, 1979 = 639, 1981 = 1,185.

With the exception of the Philippines, increases in prices were well below international oil price changes. The biggest increases were generally in bunker prices. Kerosene prices shot up, especially after the second oil crisis (1979–80), as governments were no longer able to protect the rural and poorer groups by subsidizing oil imports and providing cross-subsidies from gasoline users.

Domestic pricing policies must be responsive to a multiplicity of national policy objectives. The economic efficiency criterion provides a convenient starting point to establish rational and practical energy prices. It also provides a basis for making domestic prices more responsive to international energy price changes, so as to maximize the benefits and minimize shocks to national economies. The general principle is to follow long-run world energy price changes, and adjust local prices so as to influence or manage the domestic fuel mix to take advantage of absolute and relative international energy price shifts. This implies that short-term world fluctuations are not a good guide for domestic energy price setting. At the same time, careful analysis must be carried out by experts to attempt to forecast future international price movements on the basis of the most recent data and establish natural pricing policies as robust as possible in the face of uncertainties.

ENERGY CONSERVATION AND PRICING

Using both price and nonprice tools, demand management techniques help establish economically efficient or optimal patterns and levels of energy consumption. This may involve reducing the consumption of some forms of energy and increasing the use of others that are cheaper or more suitable.

Energy conservation is an important element of demand management and involves measures that specifically seek a deliberate reduction in the use of energy below some level that would otherwise prevail. Such reduction involves elimination of outright waste, reduction of energy-using activity, substitution of one form of energy for another, or substitution of other productive factors like capital and labour for energy.

A recent estimate (World Bank, 1983) indicates that by 1990, the developing countries can save more than 4 million barrels (bbl)/day oil equivalent or about 15 percent of total commercial energy consumption if effective conservation policies are adopted in the four key sectors outlined, although this will not be easy. Thus, inappropriate pricing of energy resources is not the only reason for inefficient energy conservation decisions. In many developing countries the lack of foreign exchange resources forces government to maintain strict import controls. Thus, it is often impossible for large energy users to import new, more energy-efficient equipment to replace that in existence, even though they are usually able to secure their share of high-cost imported fuel supplies to keep their existing fuel-inefficient equipment operating. In countries in which fuel prices are subsidized at the same time, there is little incentive for such equipment owners to press for appropriate changes in import policies.

Conservation economics

Some conservation is achieved simply by reducing or eliminating certain energy-using activities. Higher energy prices enhance these trends. Forgoing Sunday pleasure driving, using a lower thermostat setting and shutting off appliances and lighting fixtures when not directly needed are typical examples. Other conservation measures may require substitution by either capital or labour. Examples are reusing heat in industrial processing, energy-saving reductions in the weight of vehicles by better engineering or lighter materials, or the use of improved insulation.

Pricing policy also plays an important role in the substitution of some form of costly, or scarce, energy resource by another that is more readily available. This is an important conservation measure. Examples are the use of coal instead of fuel oil in heat processes, the use of natural gas instead of petroleum products for power plants where gas is plentiful compared with oil, or the use of gasohol instead of petrol for transport. In a physical sense (as measured by British Thermal Units (Btu) consumed), such substitution may not "save" energy. In an economic sense, however, such substitution may be quite sensible, given the economic scarcity values of the alternative fuels.

The pursuit of energy conservation as a goal raises the issue of up to what point the reduction of energy consumption is socially beneficial or desirable. Common sense indicates that "wasteful" energy use should be discouraged, but there is a limit beyond which conservation becomes too costly in terms of forgoing other resources or useful outputs, thereby causing more harm than good. The principal objective of a given policy should be the maximization of the welfare of a society over time.

In simple terms, the adoption of a given conservation measure is economically justified if $dB > dC1 + dC2$; where dB, $dC1$, and $dC2$ are the economic values of marginal energy-saving benefits, marginal additional input costs, and marginal reductions in consumption benefits, respectively.

This condition should be achieved over the life expectancy of the activity, implying one of expected lifetime costs, not just currently prevailing cost relationships. For example, if energy costs are expected to increase relative to other input costs or the value of output over time, greater substitution by nonenergy inputs (i.e., higher levels of energy conservation) is called for. If we introduce the time element the conservation criterion becomes:

$$\sum_{t=0}^{T} bt/(1+r)^t > \sum_{t=0}^{T} (c1t + c2t)/(1+r)^t$$

where bt, $c1t$, and $c2t$ are the respective annual energy savings, additional input costs, and losses in consumption benefits in year t, and r is the discount rate, all defined in terms of appropriate shadow prices.

CASE STUDY: CO-ORDINATED USE OF PRICING AND RELATED TOOLS FOR ENERGY CONSERVATION

In the following case study, let us consider a particular end use for energy, such as home lighting, and assume there is a choice of two distinct types of light bulbs, incandescent and fluorescent (Munasinghe and Schramm, 1983). For simplicity, we begin by assuming that both have the same economic cost, same lifetime, and provide light output of the same quality. If the fluorescent bulb uses less electrical energy than the incandescent one, then replacing the latter by the former is a conservation measure that results in an unambiguous improvement in economic as well as technical efficiency. In this case, using a fluorescent bulb instead of an incandescent lamp reduces the economic resources expended to provide the desired output, i.e., lighting. Electrical energy has been conserved, with no change in other economic costs and benefits.

Next, assume that the fluorescent bulb is more costly to install. There is a trade-off between the higher capital cost of the fluorescent lamp and the greater consumption of kWh by the incandescent bulb. The relevant data to determine whether substitution of incandescent by fluorescent bulbs is economically justified are summarized in Table 1.3. At this stage we distinguish between the economic value (or opportunity cost or shadow price, as discussed earlier) of a good or service and its market price. The former is relevant to decision making from a national perspective and the latter is more appropriate from a consumer's viewpoint.

The national cost (based on economic values) of using the incandescent and fluorescent bulbs over their two-year lifetimes are respectively:

$$ECI = 10.5 + 16 + 16/(1+r) \qquad (1)$$
$$ECF = 32 + 4.4 + 4.4/(1+r) \qquad (2)$$

Assuming an economic discount rate of $r = 0.1$, we find $ECI = 41.0 > ECF = 40.4$.

We have compared the energy cost saving of $(16 - 4.4) = 11.6$ dineros per year for two years against the increase in capital costs $(32 - 10.5) = 21.5$ dineros. We find that $(16 - 4.4) + (16 - 4.4)/(1+r) > (32 - 10.5)$. Therefore, using fluorescent light bulbs, with their associated reduction in energy consumption, will improve economic as well as technical efficiency.

Note, however, that if we use $r = 0.2$, $ECI = 39.8 < ECF = 40.1$, the conservation measure is no longer beneficial. This reduction in the relative value of conservation will always occur with increases in the discount rate, because increases in initial investment costs are traded off against the future cost savings realized by conservation. This finding has important policy implications. Energy users who confront high opportunity costs of capital (e.g., those in many developing countries) will find costly capital-intensive energy conservation measures relatively less attractive than users who have access to low-cost sources of capital. This means that economically "optimal" conservation measures may differ significantly among different countries.

Market imperfections and private consumers

So far the analysis has been based on the national viewpoint, using values for all inputs and outputs (including those for energy) reflecting economic opportunity or shadow costs. However, market prices may differ from shadow values because market imperfections, particularly in the pricing and availability of energy, abound in most countries.

To illustrate the effects of these divergencies, let us return to the simple light-bulb example. The private costs (based on market prices) of using incandescent or fluorescent lighting, respectively, are as follows:

$$PCI = 18 + 12 + 12/(1+R)$$
$$PCF = 36 + 3.3 + 3.3/(1+R)$$

At a discount rate of $R = 0.1$ (e.g., the market interest rate based on private bank rates): $PCI = 40.9 < PCF = 42.3$. This means that a rational consumer would prefer to use incandescent light bulbs, because this is the cheaper option. At any higher discount rate the advantage of the incandescent system over the fluorescent one increases further. Thus, since market prices diverge from real economic costs, consumers would make economically inefficient energy-use decisions.

Policy interactions

In addition to appropriate pricing, there is a wide variety of direct and indirect policy measures that can be taken to bring about desirable levels of energy conservation. Among them are direct regulation of energy uses, regulation of the use of energy-consuming equipment and appliances, mandatory standards, mandatory information requirements about energy consumption rates, taxes and subsidies, appropriate infrastructure investments for energy-saving facilities (e.g., better roads, railroads, marine shipping facilities), propaganda, and others.

To analyse some of the effects of such conservation-oriented policies, let us first return to the light-bulb example. As we have found, existing market prices have made it more attractive for users to opt for the incandescent light-bulb system. To resolve this difference between optimal economic and private market choices, the first option policymakers might consider could be to raise the market price of electricity from 0.3 dineros per kWh to its economic value of 0.4 dineros per kWh. We now have: PCI = 48.5 > PCF = 44.4, and rational electricity consumers will make the correct decision in favour of fluorescent lighting. In addition, setting the electricity price equal to its marginal opportunity costs will also establish electricity consumption for nonlighting purposes at optimal levels.

Suppose that public resistance or other social pressure makes it impossible to raise electricity prices. Let the economic value of an incandescent bulb be its cost of production or producer price, while the imposition of a government tax of 7.5 dineros determines the market price. Similarly, assume that an import duty of 4.0 dineros represents the difference in the c.i.f. import cost (32 dineros) and the market price of fluorescent bulbs. Instead of raising electricity prices, an alternative policy option might be to raise the tax on incandescent light bulbs to 9.5 dineros, making the market price 20 dineros. In this case, PCI = 42.9 > PCF = 42.3, which encourages the desirable consumer decision. Reducing the duty on fluorescent bulbs to 2 dineros and lowering the retail price to 34 dineros would also yield a favourable result, since now: PCI = 40.9 > PCF = 40.3.

Some combination of the tax increase and lowering of duty could also be used. From a strictly economic viewpoint and ignoring effects outside the light-bulb market, reducing the import duty would be preferable to raising the producer tax because the former action reduces the divergence between the market price and the economic opportunity cost of fluorescent bulbs, whereas the latter has the opposite effect and increases the market distortion in the price of incandescent light bulbs.

Next, assume that the tax on incandescent light bulbs cannot be increased because the legislation affects a much larger class of related products. Similarly, suppose that the import duty on fluorescent bulbs cannot be reduced because it would undercut the price of a high-cost local producer and drive him out of business. In this instance, some final options left to the energy policymaker might be to legislate that all incandescent light bulbs be replaced by fluorescent ones, or to give a direct cash subsidy to consumers who adopt the measure, or to mount a major public education and propaganda campaign to bring about the required change (Peck and Doering, 1976; Walker, 1980).

Complications

If the useful lifetimes of technological alternatives are different, then
economic comparisons become somewhat more complicated. This would
be the case in our earlier example if the lifetime of incandescent bulbs
were to be only one year, while that of fluorescent lamps might be three
years. Two alternative approaches could be used to overcome this difficulty. In the first, the investment costs of each alternative would have to
be annuitized over its lifetime at the appropriate discount rate and the
associated energy consumption and other recurrent costs for one year would
be added. Then the total costs for each option would be compared. The
second method would compare the full costs of each alternative over a
much longer period, say 20 years, including the costs of periodic replacement of worn-out equipment. The two methods should give consistent
results, assuming the same values are used for parameters such as the discount rate.

Another difficulty associated with changes in the benefits of consumption arises if either the quality or the end product of energy use is different for the two alternatives. Consider a comparison of electric versus
kerosene lamps for lighting. In addition to the differences in equipment
and fuel costs, the cost-benefit assessment of the two options should also
include a term to recognize that electricity is likely to provide lighting of
a superior quality. While the quantification, in monetary terms, of this
qualitative superiority will be difficult, one measure might be the willingness of the consumers to pay for the different forms of lighting, usually
represented by the area under the relevant demand curve.

Specific conservation measures such as rationing have a quality effect
that must be taken into account. For example, with the physical rationing
of petrol, the cost or welfare loss to the consumer due to the reduction
in the miles he can travel in his car must be added to the cost of
implementing the rationing scheme and then compared with the benefits
of reduced petrol supply. Once again, the willingness to pay of petrol users
would be the appropriate measure of the forgone consumption benefit.
However, in the long run, petrol consumption could also be reduced by
the introduction of a more fuel-efficient car engine without (perhaps)
requiring a reduction in the miles travelled. This shows that a reduction
in energy consumption does not always imply a reduction in consumption benefits; a major focus of the appropriateness of conservation policies
should be the service derived from the energy use.

Finally, the costs and benefits associated with externalities should be
included in the economic cost-benefit comparison of alternatives. For example, improvements in technical efficiency or fuel substitution measures may
give rise to pollution, as in the case of conversions from oil-burning to
coal-fired electric power plants. These additional "external" costs should
be explicitly evaluated in the analysis.

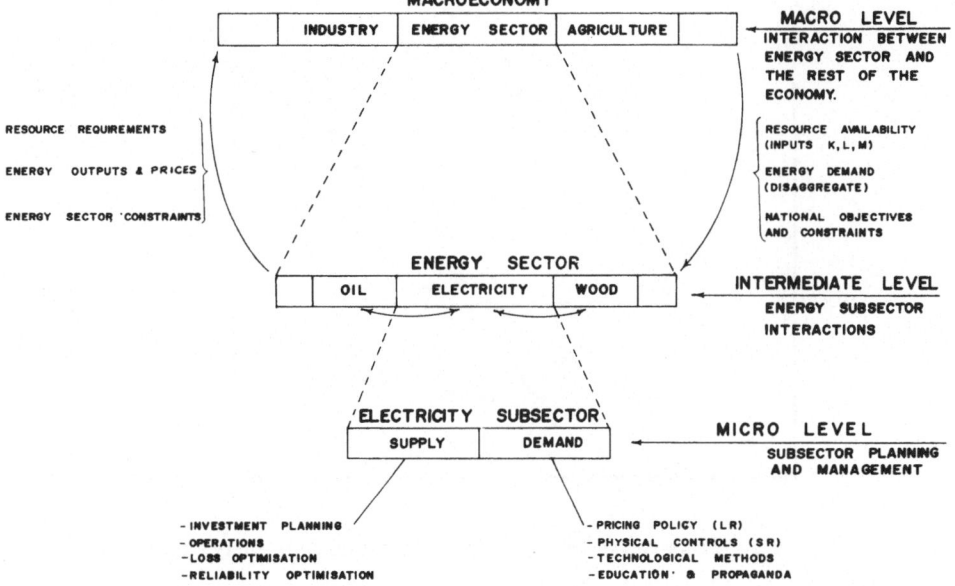

Figure 1.1 Pricing policy in relation to the hierarchy of interactions in integrated national energy planning (INEP)

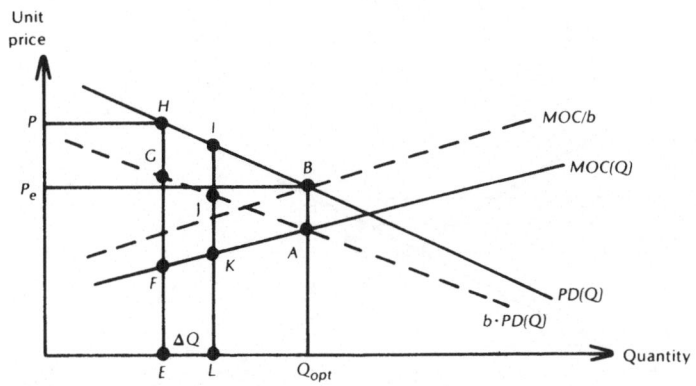

Figure 1.2 Pricing for economic efficiency using shadow prices

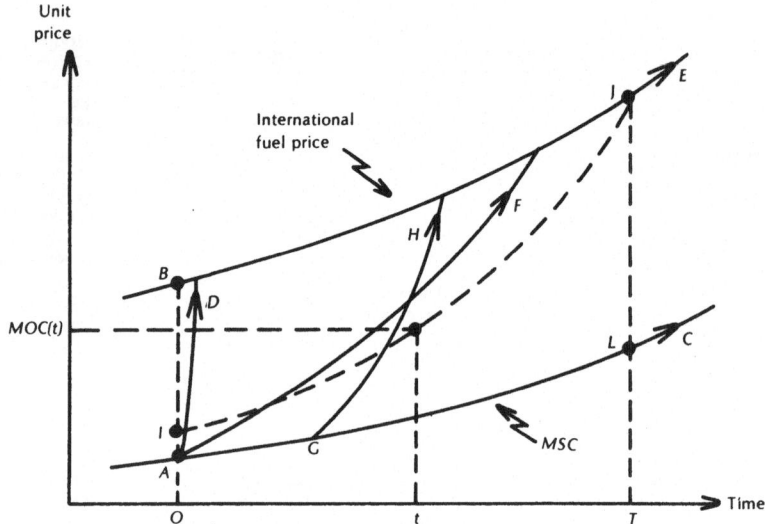

Figure 1.3 Long-run evolution of prices for domestic energy sources

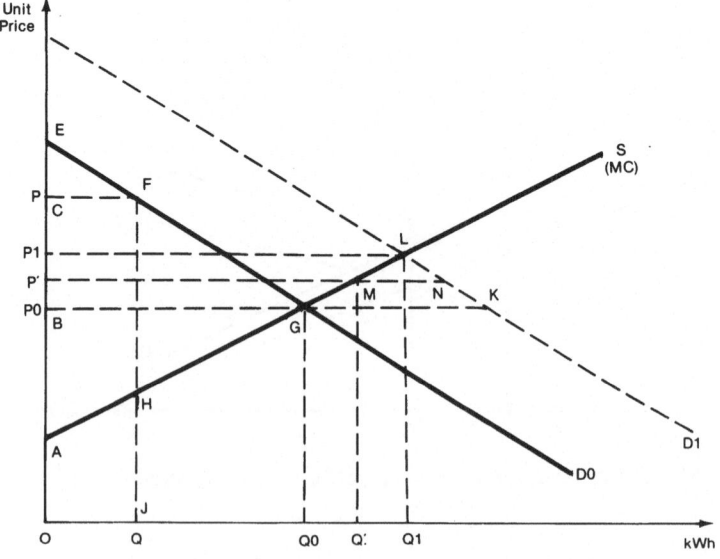

Figure 1.4 Effects of shifting demand curve

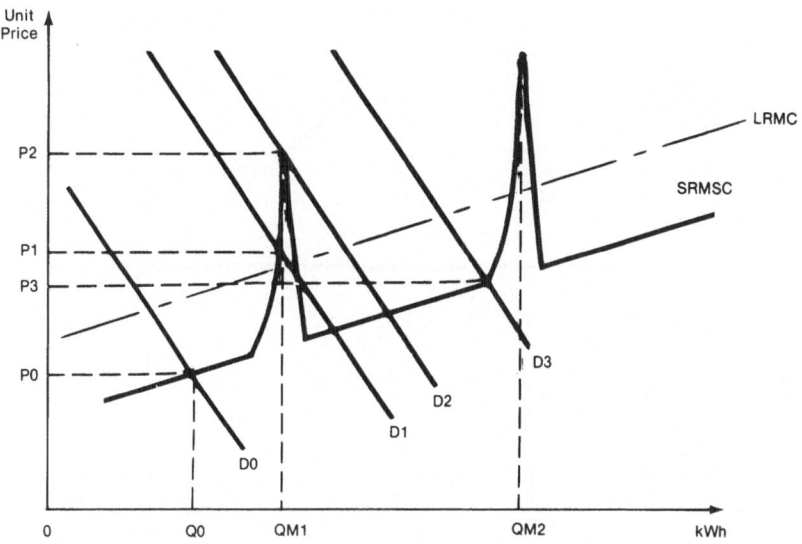

Figure 1.5 Effects of capital lumpiness

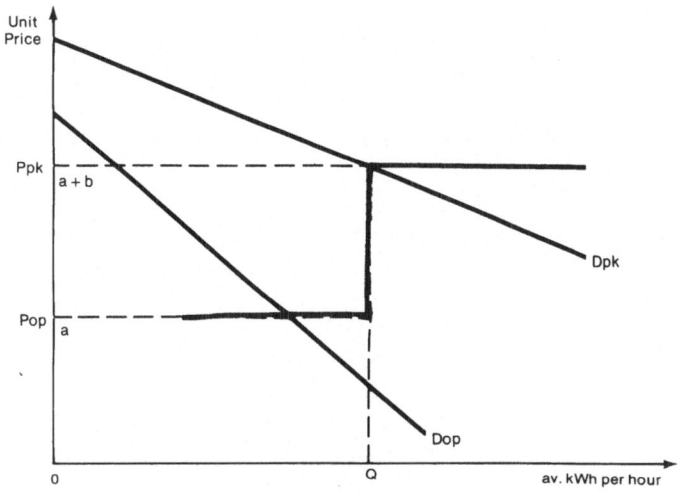

Figure 1.6 Peak and off-peak pricing

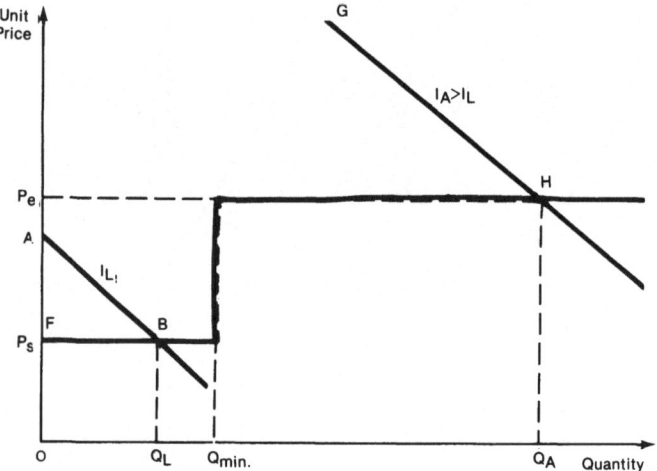

Figure 1.7 Rationale for subsidized or lifeline prices

Table 1.1 Magnitudes and types of tariffs, 1982

Country or area	Utility	Average tariff rate USc/kWh	Life-line rate	kW and kWh charges	Time-of-day pricing	Fuel cost adjustment clause
Afghanistan	DABM	2.51	Y			
Bangladesh	BPDB	5.75	Y	Y		Y
Burma	EPC	3.62	Y			
Fiji	FEA	14.83		Y	Y	
Hong Kong	CLP	8.10		Y		Y
Hong Kong	HEH	9.70		Y		Y
Indonesia	PLN	7.70	Y	Y	Y	
Lao People's Democratic Republic	EDL	1.14				
Malaysia	NEB	9.52	Y	Y		
Malaysia	SESCO	12.07		Y		
Nepal	NEC	5.10	Y	Y		
Pakistan	KESC	7.90	Y	Y		Y
Pakistan	WAPDS	4.31	Y	Y		Y
Papua New Guinea	ELCOM	15.22	Y	Y		
Philippines	MIRALCO	6.79	Y			
Philippines	NPC	5.62		Y		Y
Republic of Korea	KEPCO	8.75	Y	Y	Y	
Samoa	EPC	13.00		Y		
Singapore	PUB	8.93	Y	Y		Y
Solomon Islands	SIEA	25.40		Y	Y	
Sri Lanka	CEB	6.19	Y	Y		Y
Taiwan (Province of the People's Republic of China)	TAIPOWER	6.90	Y	Y	Y	
Thailand	EGAT	6.26		Y	Y	Y
Thailand	MEA	7.13	Y	Y	Y	
Thailand	PEA	7.04	Y	Y		
Vietnam	PC2	8.40				

Source: Munasinghe and Rungta (1984).
Y=Yes

Table 1.2 Asian developing countries: index of retail product price changes, 1973–81 (1973=100)

	Year	Motor gas (Regular)	Kerosene	Bunker C fuel oil
Burma, Rangoon	1975	173	154	263
	1977	94	140	210
	1979	104	154	211
	1981	97	144	N.A.
India, New Delhi	1975	225	174	268
	1977	200	178	310
	1979	277	228	559
	1981	156	244	632
Indonesia, Jakarta	1975	139	139	254
	1977	171	156	295
	1979	163	143	N.A.
	1981	243	219	401
Pakistan, Islamabad	1975	174	199	174
	1977	213	186	212
	1978	226	204	251
	1981	376	629	448
Philippines, Manila	1975	361	312	350
	1977	471	338	404
	1979	589	438	406
	1981	1,306	869	451
Republic of Korea, Seoul	1975	116	110	213
	1977	141	81	177
	1979	203	103	259
	1980	252	144	365
Singapore	1975	118	50	259
	1977	122	57	260
	1979	147	113	268
	1981	194	155	620
Sri Lanka, Colombo	1975	213	266	315
	1977	205	229	568
	1979	212	103	350
	1981	900	512	756
Thailand, Bangkok	1975	164	125	225
	1977	188	139	250
	1979	354	217	448
	1981	546	317	694

Source: Siddayao (1983).

Table 1.3 Physical and economic data to assess the economic efficiency of energy conservation for lighting

		Incandescent bulb	Fluorescent bulb
Installation cost (dineros)	Economic value (opportunity cost)	10.5	32
	Market price	18	36
Physical energy consumption (kWh per year during 2-year lifetime)		40	11
Value of energy consumption (dineros per year during 2-year lifetime	Economic value (marginal opportunity cost)[a]	16	4.4
	Market price[b]	12	3.3

[a] Dineros 0.4 per kWh.
[b] Dineros 0.3 per kWh.

REFERENCES

Hotelling, H. (1938). "The general welfare in relation to problems of taxation and of railway and utility rates." *Econometrica* (July).

Little, I. and J. Mirrlees (1974). *Project Appraisal and Planning for Developing Countries*. New York: Basic Books.

Mishan, E. (1976). *Cost-Benefit Analysis*. New York: Praeger Publishing Co.

Munasinghe, M. (1979). *The Economics of Power System Reliability and Planning*. Baltimore: The Johns Hopkins University Press.

Munasinghe, M. (1980a). "The costs incurred by residential electricity consumers due to power failures." *Journal of Consumer Research* (March), pp. 361–369. Also available as Reprint No. 120, The World Bank, Washington, D.C.

Munasinghe, M. (1980b). "An integrated framework for energy pricing in developing countries." *Energy Journal*, Vol. 1, No. 3 (July), pp. 1–30. Also available as Reprint No. 148, The World Bank, Washington, D.C.

Munasinghe, M. (1980c). "Integrated national energy planning in developing countries." *Natural Resources Forum*, Vol. 4, pp. 359–373. Also available as Reprint No. 165, The World Bank, Washington, D.C.

Munasinghe, M. (1980d). "A New Approach to Power System Planning," in *IEEE Transactions on Power Apparatus and Systems*, Vol. PAS-79 (May-June). Also available as Reprint No. 147, The World Bank, Washington, D.C.

Munasinghe, M. (1981). "Principles of Modern Electricity Pricing," in *Proc. IEEE*, Vol. 69 (March), pp. 332–348. Also available as Reprint No. 185, The World Bank, Washington, D.C.

Munasinghe, M. (1983). "Third World energy policies: Demand management and conservation." *Energy Policy* (March), pp. 4–18. Also available as Reprint No. 255, The World Bank, Washington, D.C.

Munasinghe, M. and M. Gellerson (1979). "Economic criteria for optimizing power system reliability levels." *The Bell Journal of Economics*, No. 10 (Spring), pp. 353–365. Also available as Reprint No. 112, The World Bank, Washington, D.C.

Munasinghe, M. and S. Rungta (1984). *Costing and Pricing Electricity in Developing Countries*. Manila: Asian Development Bank.

Munasinghe, M. and G. Schramm (1983). *Energy Economics, Demand Management and Conservation Policy*. New York: Van Nostrand Reinhold Co.

Munasinghe, M. and J. J. Warford (1982). *Electricity Pricing*. Baltimore: The Johns Hopkins University Press.

Peck, A. E. and O. C. Doering (1976). "Voluntarism and price response: consumer response to the energy shortage." *The Bell Journal of Economics*, Vol. 7 (Spring), pp. 287–292.

Samii, M. V. (1979). "Economic growth and optimal rate of oil extraction." *OPEC Review*, Vol. 3 (Autumn), pp. 16–26.

Seneca, J. J. and M. T. Taussig (1979). *Environmental Economics*. 2nd edition. Englewood Cliffs, N.J.: Prentice-Hall.

Siddayao, C. M. (1983). *Energy Conservation Policies in the Asia-Pacific Region*. Resource Systems Institute Working Paper No. WP-83-12. Honolulu: The East-West Center.

Squire, L. and H. van der Tak (1975). *Economic Analysis of Projects*. Baltimore: The Johns Hopkins University Press.

Walker, J. M. (1980). "Voluntary responses to energy conservation appeals." *Journal of Consumer Research*, Vol. 7 (June), pp. 88–92.

World Bank (1983). *The Energy Transition in Developing Countries*. Washington, D.C.

Chapter 2

SOCIO-ECONOMIC GOALS IN ENERGY PRICING POLICY: A FRAMEWORK FOR ANALYSIS

Manmohan S. Kumar

INTRODUCTION

This chapter develops a framework for analysing the role energy pricing policy can play in serving socio-economic goals in the developing countries of Asia. A number of goals relating to economic growth, industrialization, inflation, employment, and equity can be affected by energy pricing. The chapter emphasizes the interdependence between these goals and the importance of taking into account the diverse implications of any particular pricing strategy. There are, of course, a number of other powerful instruments at the disposal of governments to attain the goals. At the same time, the pricing policy is also subject to constraints relating to the cost of energy, financing requirements, and the availability of foreign exchange. One of the main objectives of this chapter will be to examine whether the goals, subject to the constraints, could be significantly promoted by managing prices of energy as a whole, and of different types of fuels.

This chapter is divided into seven sections. The first section briefly discusses the salient features of the energy sector and the economies of developing countries in Asia and the relevance of these features to the pricing strategy. The subsequent sections examine: the role of cost factors in pricing energy; the extent to which this has to be modified to take into account equity considerations; how employment opportunities might be affected by pursuing particular policies, and the possibilities of substitution between energy and other factors of production; the implications of pricing for other goals such as growth, industrialization, and trade competitiveness; some issues specifically relating to pricing of energy, in particular electricity and traditional fuels in rural areas; and the role of information and education in supplementing any price strategy.

ENERGY AND THE ECONOMY

In discussing issues of energy pricing, it is common to use, for reference, pricing strategies based on competitive market models and to apply the standard welfare framework to the analysis. It is increasingly recognized, however, that the energy situation in a given economy or region and the structure of production and trade should be the major determinants of prices. We briefly note below some well-known features of the energy sector and of the economies of some developing countries in Asia. In view of the vast diversity in the continent, there are few generalizations which can be valid. However, there are certain features which apply to a substantial number of the countries, and these are the ones which are emphasized.

The energy sector

During the last decade there has been a sharp increase in the indigenous production of commercial fuels in many developing countries in Asia. Despite this and the fact that the increase has often been greater than the increase in consumption, most countries still rely on energy imports, some very heavily so. Further, although a substantial proportion of indigenous production consists of liquid fuels, these are very unevenly distributed. These fuels, however, account for a high proportion of consumption in most countries, necessitating substantial imports. In terms of the sectoral share of commercial energy consumption, industry and transport each account for around 40 percent, with agriculture and households consuming the rest. Industry and transport differ significantly, however, in the type of fuel, with the latter relying overwhelmingly on liquid fuels.

Another feature of the energy sector is the substantial reliance by these countries on the so-called "noncommercial" fuels including firewood and crop and animal residues. These are in reality traded in the market place, although the markets for them are local or regional. As a proportion of energy from commercial sources, these fuels on average account for more than 50 percent. If animate energy is also included, they contribute up to 50 percent of total final energy consumption. There are, however, very significant differences across countries with the proportion of noncommercial to commercial energy consumption being less than 1 percent for Hong Kong and Singapore, to more than 100 percent for Burma, Sri Lanka, and Indonesia.

There are two other aspects of the energy scene which are worth noting: (1) the high elasticity of commercial energy demand with respect to economic growth, and (2) the increasing substitution of commercial fuels for noncommercial fuels. While the elasticity coefficient varies from country to country and depends critically on the time period and methodology chosen, it is generally in the region of 1.5 to 2.0. The reason for such a high value lies partly in the changing structure of the economies so that, with growth and development, the importance of energy-intensive industry and transport increases more than proportionately; partly, the reason lies in the substitution itself of commercial forms of energy for noncommercial forms as incomes increase. Both these reasons suggest that the future demand for commercial energy may turn out to be considerably greater than the current projections suggest (see World Bank, 1980, for some estimates of this).

The economy

The "stylized" facts about the structure of the economies of Asia hardly need emphasizing. While there is again enormous heterogeneity, a few common salient features which impinge directly on energy pricing are noted below. The first is the low per capita income of the countries. Development over the past three decades has led to considerable increases in aggregate real income but not to any dramatic changes in per capita incomes. In the face of this, the governments' concern that energy consumption should not impose any additional burden on the poorest people appears quite natural. The concern over the second feature, the high rates of inflation, is also understandable. In most countries inflation accelerated sharply during the past decade, and it is still on average more than 10 percent. The third feature is the role of agriculture in the economy. There has been a considerable increase in industrialization in these countries, but industry still accounts for a small proportion of the GDP. More importantly, the proportion of population employed in agriculture is still extremely high.

Another important feature is that while the growth in exports has been rapid, the propensity to import has more than kept pace with this, with the result that there are significant, and growing, problems with the availability of foreign exchange.[1] The fact, which is of relevance here, is that in recent years fuel imports have accounted for as much as a third of total merchandise imports, and almost half of merchandise export earnings. While the demand for petroleum, the main imported fuel, in the absence of any marked break in the trend, is likely to continue increasing, the prospects for increasing exports are limited. With the deceleration of growth in industrialized countries, and the Asian countries increasingly producing products which compete directly with those produced by the former, there are already difficulties in earning sufficient foreign exchange. It would hardly be an exaggeration to say that the availability of foreign exchange is the main factor constraining the further development of the economies. It will be argued here that any energy pricing strategy must keep this factor at the forefront if it is to make a contribution to socio-economic goals in anything other than the very short run.

COST OF ENERGY

Suppose initially that in setting prices of different types of fuels there were no considerations other than those relating to the costs of producing them. What prices ought to be set in that case? Even in this situation, it is argued that there is no simple rule that can be followed. This section examines the appropriateness of marginal cost pricing and the necessity of departures from this rule.

Marginal cost pricing

One criterion for pricing might be on the basis of marginal cost. These are the money outlays on factors of production required to increase output marginally. It is easy to see why this might be an attractive criterion. For satisfying optimum welfare conditions, it is required that additional consumption of a good or service should be possible at a price not greater than the additional costs necessarily incurred in producing it. The marginal cost rule distinguishes between "current" and "past" opportunity costs. It is based on the premise that once sacrifices necessary to create a durable and specific asset have been made, no further opportunity costs are incurred by its later use. Thus, because the opportunity costs have been borne in the past, no account should be taken of them in deciding current prices. This rule is derived from the criterion which stems from the analytical model of a perfectly competitive market economy. It is a property of the long-run equilibrium of the model that, given the distribution of income between consumers, no transfer of factors between users could increase the utility of one consumer without reducing that of another. The optimum conditions for welfare are fulfilled by the model. For the competitive firm it is an incidental property of the long-run equilibrium that marginal cost = average cost = price of product.

A problem which arises immediately is that the energy sector is dominated by public enterprises, which are far from perfectly competitive. Whether these are enterprises supplying coal, oil, gas, or electricity, or are refining petroleum products, a significant part of their factors of production are not perfectly divisible; they can be obtained only in large indivisible physical units, and in a durable and specific form. Further, the technically efficient production unit is large relative to the possible size of the market, and the enterprises have considerable monopoly power protected by law. In such circumstances there may be no price equal to both marginal and average cost. Pricing at marginal cost would lead to losses, and the enterprises would need subsidizing. It is worth noting that the problem of deficits need not arise only in the case of decreasing costs where the revenue yielded by marginal cost pricing will fall short of the total costs of the firm. For example, in practice, revenue requirements of the enterprises are often based on historic cost accounting, rather than on replacement costs.

It is possible that in certain cases marginal cost pricing might appear more appropriate than any other rule. For example, in the case of electricity supply, if new consumers are connected to the system, or if existing consumers increase their consumption during the system's peak, additions to generating and network capacity may be required. In this case it might be regarded appropriate that prices should equal (long-run) marginal costs so that these consumers alone bear the additional resource costs (see Turvey and Anderson, 1977). However, in general (short-run) marginal cost pricing would lead to losses which would need financing.

Before considering where the subsidies are to come from, it is useful to note that departures from perfect competition in one sector of the economy have implications for pricing in other sectors. Once one of the efficiency conditions are violated, the other general efficiency conditions are no longer desirable.[2] So even if marginal cost pricing were accepted as the appropriate criterion, if one fuel — say, kerosene — was priced below marginal cost, it would no longer be desirable from a welfare perspective, to price—say, coal—at marginal cost.

The financing of deficits

The earliest suggestion as to how the losses were to be financed focused on the use of particular types of taxes.[3] It may be argued, for example, that income taxes do not violate welfare criteria since they affect only the distribution and not the size of the national income. If such taxes were used and prices of public enterprises were equated with marginal cost, the optimum welfare conditions could be achieved. It has been recognized for a long time, however, that income taxes would themselves affect the marginal welfare conditions directly. More fundamentally, the welfare "ideal" relates to a given distribution of income, and that distribution must be altered by the proposed taxes. (An alternative might be to tax consumers of products in proportion to their consumption, but this is effectively a return to average cost pricing.) Marginal cost pricing, therefore, entails income redistribution from nonconsumers to consumers of public enterprise products. The justification for an interpersonal comparison of this kind is examined below.

It is also well recognized that the multipart tariff (e.g., in the case of electricity supply), which was intended to avoid losses, does not solve the problem either, because it requires that the fixed and the variable costs be imputed to individual consumers which, in reality, is not possible. In this case again the decision taken about the prices to be charged must involve a value judgement about the distribution of income.

The problem of financing the deficits may also be portrayed as adding an additional constraint on optimal resource allocation (see Baumol and Bradford, 1970). This requires that the price deviate systematically from marginal cost. A standard result is that, for each product, the percentage deviation of the price from marginal cost must be inversely proportional to its price elasticity of demand. The rationale for this rule is: the damage to welfare resulting from departures from marginal cost pricing will be minimized if the relative quantities of the various goods sold are kept unchanged from their marginal cost pricing proportions. It is quite difficult in practice to design a pricing structure which can ensure this. More fundamentally, this procedure continues to assume a perfectly competitive market economy and, hence, it is again open to the sort of objections previously noted (see also Siddayao, Chapter 6, of this book).

This brief discussion of the issues relating prices to marginal cost indicates the complexity of the situation and shows that one cannot get away from distributional consequences. This is particularly important since the government may want to affect distribution. In this context it is also worth noting Wiseman's remarks made more than a quarter of a century ago: "It would therefore appear that failing some universally acceptable theory of the public economy by reference to which policy could be decided, economists would find their efforts better rewarded if they ceased to seek after general pricing rules and devoted attention to the examinations of the policies actually adopted by governments, in order to discover their effects and make clear . . . the nature and consequences of the policies actually being pursued" (Wiseman, 1957).

PRICING AND EQUITY

After nearly two decades of impressive economic growth in most Asian countries, it became clear that the benefits of growth were not being shared by all sections of the society.[4] This has led to a major reformulation of the strategy for development with direct emphasis being placed on meeting the basic needs of the poor, even though this may not directly lead to high economic growth.[5] It is in this context that the goal of equity can be seen clearly. Governments have been concerned that the sharp increase in the price of oil products and the associated increases in substitute fuels should not impose too great a burden on the poor.

There are a number of different ways in which changes in fuel prices would affect household income: (1) directly, whereby there would be a change in real income (the real purchasing power for buying other goods would be changed); (2) through changes in the cost of consumption goods which directly or otherwise use energy as an input (in practice, nearly all goods); and (3) through indirect effects which alter aggregate income, the foreign exchange constraint, inflation, employment opportunities, and so on.

The essence of the equity concern is that the adverse effect on the poorer households of increases in energy prices should be minimized. Invariably the focus has been on the direct effect of fuel price increase on household budget, and we consider this first. The magnitude of this effect depends on the expenditure on fuel as a proportion of total expenditure and the elasticity of substitution between different fuels. Since prices of some fuels will increase more than others, a high elasticity of substitution may prevent too great an increase in the burden.

While the governments have been concerned with preventing too sharp an increase in the burden on the poorer consumers, this has often been interpreted in terms of measures concerned with relative income distribution, rather than with the effect on absolute real income. It is quite possible for the former to show only a marginal change but for the latter to change dramatically. It is not immediately obvious that in analysing the direct effect on equity of fuel price changes, one should be concerned with relative distribution rather than with the absolute concept. If fuels are subsidized, equity is, of course, also affected by the way the resultant losses are financed. If they are financed by increases in taxes, the type and incidence of taxes are important. If they are financed by increasing credit or borrowing, the indirect repercussion of this on the poorer people should also be taken into account. These issues are noted below, where we consider the different types of fuels used by different consumers and the total expenditure on them, the relative and absolute measures of income distribution, the effect of different modes of financing deficits on equity, and, lastly, the effect of changes in income distribution itself on the demand for energy.

Expenditure on fuel

In the household sector, energy is used largely for cooking and lighting (although in certain areas heating and cooling would also be important). There are distinct differences in the urban and rural areas. In the former, electricity is used for lighting, and a range of different fuels including natural gas, electricity, and kerosene are used for cooking. In the rural areas in most countries, although there has been a considerable increase in the provision of electricity, kerosene is probably still the most important fuel for lighting. For cooking, there is a preponderant reliance on the non-commercial fuels, especially firewood and animal residues (see Satsangi and Gautam, 1983). Although these fuels are commercially traded, the market for them is usually local, and the government cannot exercise any direct influence on their price. Therefore, the direct equity concern would be reflected mainly in the pricing of kerosene and of electricity. While the analysis here is applicable to both urban and rural areas, the pricing of electricity and firewood prices raises a number of additional issues which are examined in the "Energy Pricing in Rural Areas" section of this chapter.

There are two types of sources which provide information on the direct expenditure on fuel by different income households: the household expenditure surveys which are carried out more or less regularly by government agencies in several of the countries, and some questionnaires and specific surveys carried out by individual researchers. Not surprisingly, the former shows that the expenditure on heating and lighting by the low-income groups is considerably higher than for higher-income groups, and that

it has been increasing over time. For example, in the case of India the
National Sample Survey shows that in 1973–74 in rural areas the expenditure on fuel and light as a proportion of total expenditure varied from
around 9 percent for the lowest three expenditure classes to around 4 percent for the top three classes (see Bhatia, Chapter 5, of this book). As
a proportion of nonfood expenditure, fuel and light claimed more than
50 percent for the lowest three classes and only around 10 percent for the
top three. The situation in the urban areas was similar. Although no
systematic recent data are available, it appears that there has been a sharp
increase, rising to more than 20 percent, in the proportion of total expenditure on fuel and light by the poorest people (see Satsangi and Gautam,
1983). The findings by individual researchers are equally disturbing. For
example, Eckholm (1980), after an extensive survey, noted that with the
sharp increase in kerosene prices in most countries, the prices of firewood
also increased dramatically so that "some manual labourers had to spend
nearly a quarter of their total income on firewood" (p. 64). A number
of other field studies undertaken in recent years reach similar conclusions
(see Smil and Knowland, 1980). Although these studies are not substitutes
for the country-wide surveys carried out using stratified samples by the
national agencies, they are invaluable in highlighting the effect on the
poorest consumers. As such, more resources should be devoted to them
to obtain further detailed information on the economic welfare of these
groups of individuals.

Budget constraint and income distribution

It is worth noting in this context that in the standard analysis of the budget
constraint facing the consumer, there is a preponderant emphasis on consumer preferences; but preferences assume a degree of substitutability. For
example, with a linear budget constraint and expenditure y, the following
has to be satisfied:

$$y \geq \sum_{i=1}^{n} p_i q_i$$

where P and q denote the prices and quantities of goods. With two categories — say, fuel and food — the situation in Figure 2.1 prevails. This
type of analysis is likely to be inappropriate for the sort of situation facing the poorest people, where there is a basic survival constraint. If we
denote by q_1^{min} and q_2^{min} the minimum quantities of food and fuel
necessary for survival, the choice is restricted to the triangle ABC. For
a household with a budget as low as $y = p_1 q_1^{min} + p_2 q_2^{min}$ there is no
choice; it must buy at A or cease to exist (cf. Deaton and Muellbauer,
1980, Chapter 1).

    For a large number of households the budget indeed offers very little
choice and, in these cases, any significant increase in the price of fuel is

likely to have notable consequences. This suggests that a measure of the real income of households based on absolute level is likely to be more appropriate when examining price changes than a measure based on relative income. Most of the theoretical analysis and empirical data relate to the relative income distribution and we consider this first.

On *a priori* grounds one would expect changes in fuel prices alone to directly exert only a limited influence on the distribution of real income.[6] There are a large number of other more important factors which determine the distribution. These include the ownership of assets (see Chenery *et al.*, 1974), differences in the level of education, differences in labour productivity and terms of trade between agricultural and nonagricultural activities, inflow of foreign capital, and pattern of taxes and government spending (see Ahluwalia and Carter, 1979). Whatever the cause, there is clear evidence that income distribution is highly skewed and that, in terms of nominal income, inequality has worsened in the past decade. The inequality measures are all based on the Lorenz curve. This is in the sense that the income distribution depicted by the Lorenz curve is used to construct an index of inequality such as the Gini coefficient, the Atkinson and the Theil indices, and many others. Also, like the Lorenz curve, they are mean independent. If everyone's income changes by the same constant percentage, relative inequality is unchanged (see Atkinson, 1975). A number of studies indicate that taking into account both the direct effects and the indirect effects of increases in fuel prices has no significantly adverse effects on the distribution. Hughes (1983a and 1983b) notes, for example, that substantial, hypothetical increases in the price of kerosene in Thailand and Tunisia would lead to only marginal worsening of the income distribution.

Changes in income distribution gauged by measures of inequality tell only part of the story. As has often been pointed out, the absolute income measures, which focus on changes in the income of groups of individuals without paying any attention to the rest of the distribution, may tell a very different story.[7] Fields (1980) notes, for example, that in the case of India in the 1960s, while there was a notable improvement in income distribution, there was a sharp decline in the real income of the poorest households. There do not seem to be any systematic data available to indicate the precise effect which increases in the fuel prices have had, or could have in the future on the poorest households. But one can say as a rough estimate that, if the expenditure by poor people on fuel and light accounts for around 20 percent of total expenditure, a 50 percent increase in the price of energy (with inelastic demand) would lead to a 10 percent reduction in their real income. A similar price increase for a high-income household with only 4 percent of expenditure on fuel and light would lead to a diminution of only 2 percent in real income. In this case, price increases lead to a highly adverse effect on the absolute real income of the poorest households, but it may appear marginal in terms of change in nominal income distribution. Any proposed significant increases in prices must take this into account. At the same time it is imperative that much more information is acquired on the direct consumption of energy by the poorest households than is at present available.

Financing and other constraints

Obviously the matter of equity does not stop there. If household fuels are to be subsidized, it may be very difficult in practice to discriminate between consumers in different income classes. Past experience in subsidizing other commodities shows that the benefits are seldom confined to the intended recipients. Even in the case of electricity, for which it may be thought possible to confine subsidies to the poorer people, there are likely to be unintended consequences. For example, if tariffs are lowered for consumption below a certain level, richer households consuming electricity below this level will automatically benefit. A general subsidy for kerosene may lead to its being substituted in part for gasoline for the private transport of richer households. If it is argued that these subsidies will lead to a smaller fall in the absolute level of real income of poorer households and this is desirable in itself, then there is the obvious question of how the subsidies are to be financed. If they are financed by general taxes on commodities, this may in fact lead to redistribution from poor to rich.[8] This is because poorer households that consume no electricity, or few petroleum products, end up contributing proportionately more. If the government does not raise taxes but prints money, this may have inflationary consequences which are regressive.

It is also possible to cross-subsidize fuels, and this is not an uncommon practice. In other words, government could increase the price of, say, gasoline more than the amount warranted by costs and subsidize kerosene with the proceeds. If gasoline were only used by the high-income households for private cars, this might be a satisfactory solution. However, since public road transport used by poorer people also uses gasoline, this would affect transportation costs and hence prices. It may be possible to provide gasoline on subsidized terms to transport, but this may conflict with the goal of encouraging rail transport. (A number of other issues concerning transport are noted in the "Employment" section of this chapter.)

As some observers have suggested, a different consideration is that price increases, especially for indigenous fuels, may lead the utilities producing them to become less efficient or to slow down their conservation efforts (see Fallen-Bailey and Byer, 1979). Most utilities have set financial targets in terms of covering their average costs and obtaining a satisfactory return. Frequently, these targets are not met, and there are demands from the utilities to allow increases in prices of their products rather than attempt to reduce costs. The government can adopt a different strategy by not allowing increases in price as such but by taxing the fuel so that its revenue goal is satisfied and the utilities still have the incentives for cost minimization. As noted earlier, since in most countries utilities are public corporations, there are no distributional considerations relating to this, whereas, there would be if utilities were privately owned and the reduced profits had to be borne by private individuals in the absence of any price increases.

It is also the case that prices are important elements in the energy demand policies. Even though in the short run demand may be inelastic, in the long run in order to reduce growth in demand for, say, kerosene, the government may increase its price. There are several considerations which are relevant here: to the extent that prices of competing fuels also rise, the substitution for kerosene may be limited. On the other hand, if the government is also concerned with containing the demand for firewood, the relative price of kerosene may actually have to be reduced.

It may also be possible to make electricity more competitive, but this assumes unrealistically that all the desired supply would be forthcoming. In any case, it is unlikely to be of any help to consumers who have no electricity supply and are unlikely to obtain it in the near future. Another consideration is that in the short run, while there may be some increase in efficiency with which the fuel is utilized, it is unlikely to be very much. This is because the efficiency depends on the type of equipment being used and availability of alternative, more efficient equipment. For example, it is unlikely that there will be any significant change in stoves for cooking and lamps for lighting in the short run.

Income distribution and energy demand

We have noted earlier how changes in fuel prices may affect the distribution of income. Suppose now that income distribution is changed, largely by factors other than those relating to the price of energy. An interesting question then is the following: Would a more egalitarian distribution have any appreciable effect on the demand for energy? There is considerable evidence that politically feasible changes in income distribution do have a noticeable effect on the structure and performance of the economy. For example, Paukert et al. (1981) found that in three of the four countries they examined, a progressively hypothetical redistribution of income in favour of the lower classes would lead to an increase in the level of employment and also to a certain, but less strong, increase in the level of output.[9] This was due to an immediate consequence of the income redistribution, namely, a reduction in the income-saving ratio and a shift in the pattern of demand in favour of more labour-intensive products — in particular, in favour of agriculture and food products. Although Paukert et al. (1981) did not examine the effect on energy demand, these shifts suggest there is likely to be a change in that. This is because the energy intensiveness of the products for which demand increases, per unit of value added, is likely to be considerably lower.

Direct estimates of the changes in energy demand following a redistribution strategy are obtained by Behrens (1984). He examined the consequences for the Brazilian economy of a hypothetically more equal distribution. The result was as for the three Asian countries studied by Paukert *et al.* — i.e., an increase in the demand for labour-intensive products, an increase in employment, and an increase in output — but the requirement for total energy also increased slightly (relative to the increase in income). However, it is likely that this result overestimates the change in total energy requirements since it is considerably affected by the high growth pattern of wood and charcoal consumption. The model used in the simulation has fixed consumption structures and so does not allow income-induced substitution effects between fuels. A more egalitarian society may be expected to bring about a substitution of charcoal and firewood by more energy-efficient fuels, such as LPG and kerosene in cooking and electricity for lighting. There is a need for further research in this area before any firm conclusions can be reached.

EMPLOYMENT

Changes in the price of energy, by directly affecting the cost of production and by affecting the choice of techniques, may exercise a considerable influence on employment opportunities. With the share of unemployment and underemployment in several Asian countries at present around 30 percent of the total labour force and with increasing urbanization, energy pricing policies assume significance for employment creation. Further, as considerable evidence shows, there is a high association between the degree of unemployment and poverty among low-income households (see Visaria, 1980), so any pricing strategy which can increase employment may also satisfy the basic needs and equity objectives.

There are a number of issues relating to employment with regard to the manipulation of energy prices as a whole and of the prices of different fuels. One issue is the extent to which increases in energy prices may lead to increases in costs, decreases in profit margins, and also decreases in investment and employment. Another is the extent to which relative price changes of fuels can be used to encourage the substitution of less energy-intensive and more labour-intensive techniques of production. This links directly with the literature on the appropriate technology for developing countries. The effect on employment will be through not only the change in technique but also the effect on the overall constraints facing the economy, in particular the balance-of-payments constraints. The effect of changing prices on the supply industries and on their employment should also be considered. Finally, the part played by expectations concerning price changes and the risk entailed in introducing new techniques are likely to be important in affecting the employment goals.

The discussion below first notes the direct effect of increases in oil prices in the past decade. This is followed by a discussion of the appropriate technology and substitution possibilities between fuels and between energy and nonenergy inputs in industry, agriculture, and transport; the relationship between risk and innovation; and the possibility of a trade-off between equity and employment.

The exogenous price increase

It is generally argued that the role of demand factors in generating employment in developing countries is rather limited because of the severe structural or supply constraints. As discussed above, however, changes in income distribution, by increasing demand for labour-intensive products, do lead to an increase in employment. It is also worth noting that the increases in the price of oil during the past decade have led to severe unemployment problems, in part by reducing aggregate demand.

The immediate effect of the oil price increases was to lead to a sharp deterioration in the terms of trade of oil-importing countries — whether developing or developed — and a sharp increase in the current account deficits of most developing countries. The transfer of purchasing power to the oil-exporting countries, whose marginal propensity to consume out of the windfall gains was considerably lower, simultaneously led to marked deflationary pressures in the economies of oil-importing countries (see Ostry *et al.*, 1982). A large part of the price increase was passed on to the final consumers but, at the same time, governments worried by the increase in inflation pursued restrictive monetary and fiscal policies which led to a further contraction in the activities of both industrial and developing countries; the contractions mutually reinforced each other.[10] This led to a further increase in unemployment in the developing countries and a sharp reduction in their growth.

The choice of techniques

This is far from denying that the structural characteristics of the economies lie behind the continuing high unemployment rate and have to be tackled to make a dent in this. There are two sets of interrelated factors which have received the most attention. The first set includes factors relating to the operation of the labour markets in developing countries. These markets are highly segmented, and government policies are considered to have led to severe distortions and imperfections. However, it is generally recognized that while these may have led to some adverse effects on employment growth, they are not likely to have been directly dominant (see Squire, 1981). The major factor is the inappropriateness of the technology — whether it is in industry, in agriculture, or in the services. The technology is regarded as being too capital-intensive, and it leads to production of products which in turn are capital intensive.

In industry, the policy options would be straightforward if capital-intensive technology was at the same time energy-intensive, and labour-intensive technology, while economizing on capital, also economized on energy. In this case a substitution for the capital-intensive technology would lead to two types of benefits. *Ceteris paribus,* employment per unit of value added would increase, and energy consumption would decrease. It may be thought that secular increases in the prices of industrial fuels, in addition to reflecting higher costs, may also lead to switching to more appropriate production technology, but this may not turn out to be so. Such increases would, in the first instance, lead to an increase in the cost of production — the magnitude of this being dependent on the proportion of direct costs accounted for by energy. Depending on the market conditions facing the producer, a number of different responses are possible within the two polar cases: (1) all the increases in costs are passed on to the final consumer; or (2) all the increases in costs are absorbed by the producer. In the first case, the increase in prices of final products may lead to considerable reduction in demand and output, with adverse consequences for employment, but the effect on the producer is likely to be small. It is in the second case that profitability might be reduced considerably, and this is when there will be an incentive for the producer to search for ways of reducing the costs.[11] In the short run, it may simply be in terms of utilizing energy more efficiently. If the energy costs are a small proportion of total costs, this may be the end of the matter. However, if the cost increases are substantial and energy accounts for a significant proportion of total costs, in the long run the producer may consider using different types of techniques. Still maintaining the assumption that these techniques are available and are labour-intensive, the producer will take into account a number of other considerations before making the change, such as the expectations regarding the future prices of energy, the financing requirements for the new technique, the conditions in the labour market, the wage rate, and other operating costs (see also Saicheua, 1984).

Matters are more complicated once we allow relative prices of fuels to vary. Suppose the government raises the relative price of imported diesel fuel compared with the price of indigenous coal. The same sort of considerations as above apply, but now there is even less certainty that a switch to technology using coal would necessarily directly lead to any greater employment. There may be some indirect benefits for employment which operate through a reduction in demand for imported oil and some relaxation in the foreign exchange constraint.

There is very limited evidence on the possibility of substitution between energy and labour in the developing countries. Some case studies for individual industries have found that labour-intensive techniques can also be fuel-saving (for example, see Stewart, 1981), but it is impossible to tell the extent to which these can be generalized to the whole of the industrial sector. However, there is considerable evidence of substitution possibilities for the aggregate industrial sector in the advanced countries. For example, OECD (1982), in analysing econometric studies using production functions, concluded that for both industrial aggregate and industrial subsectors "the most common configuration of interfactor relationships emerging from empirical studies is that energy and labour, energy and materials . . . are substitutes" (p. 40).[12] The elasticity of substitution ranges widely from 0.6 to 2.2 depending on the choice of data, the estimation period, and computational procedures. At the same time, there is considerable evidence on interfuel substitution. In view of the crucial importance of the magnitude of these substitution possibilities, it is imperative that detailed analysis be undertaken for the Asian developing countries.

In agriculture, the main constraints on output and employment are said to lie in the highly unequal distribution of land (see Cline, 1977; and Berry and Cline, 1979). But, given this, it is plausible to argue that pricing of energy would have some effect. In most countries, agriculture uses small but increasing quantities of commercial energy, and the possibilities of interfuel substitution are limited. However, there can be significant effects through the substitution between commercial energy and human and animal labour. Commercial energy is used in the mechanization of irrigation and of ploughing. In irrigation, electric or diesel-powered pumps can be substituted for animal-operated devices such as water lifts with human supervision. In ploughing, tractors can replace ploughs drawn by bullocks or other animals. In the case of several Southeast Asian countries, small power tillers may serve the purpose of both ploughing and irrigation (see Jequier, 1979). Suppose that prices of two main fuels used in agriculture — diesel and electricity — are subsidized, leading to an increase in mechanization. Initially, since the productivity of both pumps and tractors is higher than that of animals, mechanization will lower the labour time required to do a certain task. Since the machinery requires supervision, as do the animals, the primary effect of mechanization must be to increase productivity and to reduce employment per task of the supervisor, thus reducing employment per unit of output of both pumps and tractors (cf. Desai, 1981).

The introduction of pumps in a country like India is likely to lead to a significant increase in the quantity of water. The draught-power applications may also increase output in areas where tractors are introduced, but they are unlikely to lead to any major extension in the margin of cultivation. The return to water inputs is likely to be very high in arid parts, whereas the return to marginal increases in draught power is likely to be limited. This suggests that the initial fall in employment from the use of pumps can be more than offset by labour employed in harvesting and processing the additional agricultural output. It is probably true that increases in the amount of draught power have much smaller secondary employment effects which can balance the initial fall in employment. In general, however, it is not obvious that subsidization of commercial energy by itself would necessarily lead to a fall in employment. There are some aspects of interfuel substitution which may also exert an influence, and these are examined in the "Energy Pricing in Rural Areas" section of this chapter.

Next consider the possible effect on employment of changes in the price of transport fuels. There are three aspects of this. First, in most countries transport accounts for a very high proportion of (imported) liquid fuels. Any pricing policy that leads to a more efficient utilization of fuels would save foreign exchange, which would have beneficial effects on output and employment.[13] It is generally acknowledged that the energy efficiency of road transport is lower than that of rail transport. A switch from one to the other may thus have some benefits. In the case of private cars with very high income elasticity of demand, taxes on gasoline while serving equity goals may also conserve energy. Another consideration is that in many countries there is a growing imbalance between domestic refinery supply and demand for petroleum products, resulting in additional net import requirements. Appropriate taxation policies may also lead to some benefits in this. Second, if energy pricing can improve the efficiency of goods transportation, this may lead directly to considerable improvement in capacity utilization and employment in the productive sectors. As in the case of industry, however, the response to increases in transport fuel costs depends on the structure of markets. If prices of transport services cannot be increased, the resultant decrease in profits may exert pressure for a more efficient service to the consumer. It seems, however, that the inadequacy of transport service is due to structural and organizational factors which are unlikely to respond, at least in the short run, to changes in fuel prices. Third, transport pricing may influence the locational decisions of firms. This is likely to have some effects on distribution of employment in different regions but probably would have no marked effect on aggregate employment.

So far we have implicitly assumed that it is the processes of production which change, with the composition of final products remaining the same. It is at least arguable that changes in energy prices may lead to changes in the type of products being produced, which may then in turn entail changes in the process of production and in employment.[14] Take, for example, the production of steel, which is both highly capital and energy intensive and in which a number of Asian developing countries have a sizeable capacity. It has often been argued that, in view of its high capital intensity, it is inappropriate for these countries. Now its high energy intensity would seem to strengthen the argument.

The theoretical basis for this is that the countries should specialize in labour-intensive (and energy-conserving) products in which they have comparative advantage and trade them for capital-intensive products. This could both increase their employment and reduce the demand for energy. In reality it would appear that the options are much more limited, and this is where conflicts of goals become so apparent. For example, the countries may not want to be constrained by a particular configuration of factor endowments which is as much due to accidents of history as to anything else. Acting according to the existing endowments may make the future production structures even less acceptable. This is quite apart from considerations relating to the fact that these products, if not produced domestically, would have to be imported, probably at even higher shadow costs; reliance on foreign supplies may also be undesirable for various strategic reasons.

It may also be argued that the energy intensity by itself is not the only criterion for evaluating the comparative advantage. The fuel composition is also important. If certain fuels are produced domestically but cannot be exported, comparative advantage may still lie in producing products which utilize these fuels intensively. Take, for example, the production of steel in India. This uses significant quantities of coal which has low opportunity cost. In this case it may be optimal to produce steel and even to export it. Obviously this would not continue to be the case if the indigenous supplies were insufficient, and the fuel had to be imported.

Risk and innovation

There are a number of facets of risk and uncertainty which can be important. The first is that whether changes in relative fuel prices lead to the production of new products or innovation of new processes, the likelihood of economic and technological risk would be substantial. In such a case it would not be sufficient for the government just to alter the relative prices, but it may also have to provide information and guidance and be ready to bear part of the risks. This also assumes that the market mechanism would lead to sufficient innovations in the first place. This is most unlikely to be the case. Rather, the government may itself have to provide research and development funds and various other facilities and incentives for this.

A second aspect is that any significant change in output composition and techniques of production would require that the economic infrastructure and the supporting services be adapted to it. This might require, for example, different types of transport facilities in industry and storage requirements in agriculture. It is also important that uncertainty relating to the price of fuels is reduced as much as possible. With imported petroleum products, there is little which can be done except in the very short run. However, with domestically produced fuels it should be possible to reduce uncertainty in the future path of prices.

Equity and employment

Are there likely to be any conflicts between these two goals? In general, it may appear not, since employment-generating policy would lead to more equitable distribution, while greater equity may itself lead to more employment. In some instances there may be a trade-off between the two, although its magnitude is unlikely to be large. Consider, for example, an increase in the price of kerosene. While this would have an adverse effect on the purchasing power of the poorest people, it could also lead to greater demand for the substitute — firewood. The gathering of firewood and the reforestation schemes are highly labour-intensive and may have some beneficial employment consequences. But clearly an increase in reforestation programmes does not have to depend on inequitable price increases; the government can institute this as an independent and necessary strategy, although in the case of the private farmer or trader the incentive may be provided only with the price increase which is determined in part by the price of the substitutes.

More significant conflicts are likely to arise through the efficiency with which fuels are utilized and their conservation. If fuels used in manufacturing low-income products are subsidized in order to keep costs and prices low and, hence, equitable in some sense, the net effect might simply be to let inefficient usage of energy continue. This would have adverse implications for the foreign exchange situation and future availability of energy, with some adverse effects on employment. Another instance arises in the context of equity, not between individuals but between different regions in a country. Equity and employment considerations might dictate that industry be located in areas away from raw material sources and main markets. If there is significant transportation involved, this may lead to wastage of energy and subsequent adverse effects on employment.

GROWTH AND OTHER GOALS

Despite the dissatisfaction with its distributional aspects, economic growth still remains the major objective for most developing countries in Asia. In the past, the constraints were thought to lie in a diverse range of economic and social spheres. Since the early 1970s, energy has come to be seen as an additional major constraint. The problem arises essentially through the balance of payments. In order to maintain growth, increasing amounts of energy have to be imported at much higher real prices than in the past. With export earnings expanding at a slower rate, this means that the proportion of earnings available for nonenergy imports necessary for production is reduced correspondingly.

This then has detrimental effects on capacity utilization, on productivity, and on growth. The issue for energy pricing then becomes the extent to which it can lead to energy conservation, or efficient utilization of energy. In the longer run, a more fundamental issue is the extent to which it can lead to changes in the structure of the economy which are more in keeping with the energy constraints. A number of authors have argued that government should also try to contain the inflationary consequences of the oil price increases because of their effect on growth and on equity.

We first examine the goal of industrialization. This is followed by a discussion of how changes in prices may affect exports and, hence, change the foreign exchange constraint from the supply side. Lastly, the manipulation of prices to contain inflationary pressures is noted.

Industrialization

The goal of rapid industrialization is subscribed to by almost all the developing countries. Industrial energy demand is a function of the absolute size of the industrial sector, the structure of output, and its energy intensity. During the past decade or so, industrial production and industrial energy demand have been growing faster than output and energy consumption in other sectors of the economies. Further, in many countries there has been an increasing trend towards heavy industry, which is highly energy intensive.

Most of the energy consumption in manufacturing occurs in the following five industries: iron and steel, other basic metals (in particular, aluminium and copper), chemicals (fertilizers), cement, and pulp and paper. The main point to note is that these all have significant energy-saving potential. Since energy costs account for a high proportion of total costs, ranging from 15 to 25 percent, an increase in the price of fuels could lead to considerable increases in total costs. The first four of these industries are generally under public ownership, and it may be possible not to allow any product price increases. This may lead to savings in energy in the short run due to better housekeeping and more efficient utilization with the existing equipment. In the long run, it could lead to changes in the techniques of production. It is worth pointing out that the more rapid the growth of the industrial sector, the easier it would be to install the latest vintages of machinery. The financial constraints are likely to be less binding, and the new capacity would be expected to utilize techniques which are more energy efficient. Another aspect of this is that there is some evidence that choice of fuel itself influences energy efficiency. For example, in cement manufacturing, kilns that use primarily coal have higher energy intensities than most oil- or gas-burning kilns.

A substantial part of the manufacturing output is still accounted for by textiles and food manufactures. These activities are less energy-intensive, and it may be thought the governments should encourage their expansion. If this were the only consideration, it might be quite possible to say electricity should be provided to them at subsidized rates which are lower than those charged to the energy-intensive industries. In this context, it should be noted that in most countries, the so-called "small-scale industries" sector accounts for a sizeable proportion of total output. Although the efficiency with which energy is utilized in this sector is probably lower than in the "organized" sector, the energy requirement per unit of value added is still relatively quite small. This sector already receives considerable state encouragement because of its employment-generating benefits. From an energy point of view also, it merits special treatment.

It hardly needs emphasizing that changes in energy prices are not the

only, or even the most appropriate, way of influencing industrial structure. Neither, of course, is energy efficiency the sole factor in technological
choice. An interesting illustration of this is provided in the manufacture
of steel. Despite the sharp increases in price of coal and electricity, the
physical scarcity of scrap and coking coal has led to an increasing reliance
on currently more energy-intensive (but with more efficient utilization),
fuel-injected processes. Another illustration is in cement production, where
the dry heat process in conjunction with suspension preheating is much
more energy efficient than the old wet process, but high capital costs have
constrained plant conversions in several countries.

It remains true that energy intensiveness in the manufacturing sectors,
except where the capital stock is relatively new, is usually greater than in
similar industries of developed countries. Factors which are seen as reducing
industry's energy efficiency include capacity underutilization, poor thermal insulation, and very high wastage of heat and gas. It is quite likely
that appropriate energy price changes could exercise some beneficial effects
in this respect.

Lastly, in pursuing the goal of industrialization, considerable attention
should be paid not just to future energy prices but also to future prices
of final products. If a number of countries pursue similar strategies, this
could lead to an excess supply of particular products with potentially
serious implications for the terms of trade. For example, a number of countries have plans for converting refineries to produce gasoline for which
there is an excess demand, rather than diesel. However, if this is put into
practice, it could very likely lead to problems of excess supply.

Exports and the foreign exchange constraint

When examining the effect of increases in energy prices on industrial costs,
obviously all stages of production have to be taken into account through
the input-output framework. To what extent are the total increases in costs
due to energy likely to reduce external competitiveness? As noted in the
previous section, a great deal depends on the market structure. In the market
environment where output prices are determined by the producer, it is likely
that some sort of cost-plus rule is followed. According to this, the final
price of the product is based on a measure of the average cost of the product
(where average cost is calculated for some normal level of capacity utilization) and a mark-up to reflect profit rate. In this case it is likely that increases
in energy prices would lead to some increases in the prices of final products. The effect of this on the export earnings depends on the price
elasticity of demand. In the short run, this is probably quite small. However, in the long run, there may be some adverse effects on export revenues.

Similar sorts of considerations apply to the exports of agricultural goods.
The direct increases in the costs of energy would mean that the producers
using machinery, tubewells, tractors, etc., would have to pay more. There
would also be indirect effects occurring through changes in the cost and,

hence, price of fertilizers and other chemicals, storage, and transportation. But in the case of agricultural exports, it is likely that the demand is highly price elastic so that increases in costs reduce farmers' margins with detrimental effects on subsequent investment. This may conceivably lead to a switch to activities which require less energy but which may not be as important from the viewpoint of earning foreign exchange.

There would seem to be a trade-off between the foreign exchange earnings expended on importing fuels and exports. Increasing energy prices may reduce growth in the former, but at the same time they may also have an adverse effect on export earnings. Keeping energy prices low would allow exports to continue expanding and would also increase relatively more the demand growth for imported fuels. The implication of this for actual pricing strategy depends critically on the magnitude of the two types of responses, as well as on other considerations noted earlier about financing requirements, equity, and employment.

All this implicitly focuses on one country. What happens if competitors also follow the same pricing strategy and raise their prices? Superficially it may appear that this would have no effect on the home country's earnings, but this would only follow if the price and income elasticities and the techniques of production were identical in the competitor countries. It would also be necessary to assume that there were no other differences in export taxes and subsidies. In practice this is unlikely to be the case, so that there may well be differential effects on export performance.

Inflation

It has often been argued that increases in energy prices directly and indirectly lead to increases in the overall inflation rate. This then leads to demand for higher wages, leading to further increases in prices. The increase in inflation has regressive consequences and can have adverse effects on output, trade competitiveness, and employment. Although all this is probably true, the government budget constraint may leave no option. In such a case, it may be possible to tax heavily those fuels which have high income elasticity of demand and which do not affect production costs unduly. This may still require inequitable price increases, but the alternatives may be even less acceptable. If the government prints money, this may lead to even higher inflation in the long run. Another option would be to borrow money. Since the domestic sources may well not be sufficient, recourse may be needed to international markets. The opportunity cost of such an action may far outweigh the inflationary consequences of the original price increase.

ENERGY PRICING IN RURAL AREAS

In the previous sections, we have noted the problems of pricing of commercial fuels which are the substitutes for the traditional fuels, including firewood, charcoal, and animal and crop residues. Demand for these fuels is heaviest in the rural areas and is dominated by household use. Rural areas do rely to an increasing degree, however, on commercial fuels including petroleum products, and on electricity. In most countries rural electrification has been increasing at a rapid pace, and there are some interesting equity and employment considerations relating to the pricing of electricity in rural areas. These are examined next, followed by a discussion of some issues concerning the pricing of traditional fuels.

Electricity prices in rural areas

Electrification of rural areas has been a major goal of most governments. Although in aggregate terms its contribution to energy requirements in rural areas is still small, it plays a crucial role in agricultural and industrial uses and in households. The connection of a village to the electricity grid does not mean that its total population has access to it. In most countries, for example, it is estimated that in the villages connected to the grid less than a quarter of the houses actually have electricity connection (Cecelski and Glatt, 1982). Since the fixed costs of obtaining the connection are high, the households that do have electricity are generally much more well off than others without electricity. Electricity is also used extensively for irrigation purposes through the powering of irrigation pumps. The major alternative for irrigation purposes is the diesel pump, though biogas is also widely used. Electricity is also used in cottage industries, but its use here is very small.

The price of electricity may be expected to be higher in rural areas than in urban areas on the basis of cost considerations alone. Marginal costs are higher for serving these areas due to the dispersed nature of demand. It is generally accepted, however, that prices should be below costs in the early years of electrification because costs are very high before demand has developed to a reasonable load factor. A more important consideration is that the provision of cheap electricity by subsidization is regarded as necessary to promote its use. In a number of countries tariffs are still lower for certain activities, such as irrigation, than for others. In terms of foreign exchange considerations also, it may appear particularly worthwhile to price electricity competitively, so that it can replace kerosene for lighting, and diesel oil for motive power. This assumes correctly that the supply of electricity is not based on oil imports as well. In the case of Pakistan, for example, central station electricity is generated using cheap hydropower and natural gas with few alternative uses; in India, it is generated mainly using indigenous coal and hydropower. But since foreign exchange savings do not have infinite value, differences in the efficiency of burning fossil fuels should also be taken into account.

The case for subsidization of electricity is said to rest on equity grounds and on the grounds that users will make different decisions about production and location of enterprises on the basis of the price of electricity.

With regard to direct effects, it is not obvious that this would have equitable consequences. Benefits from subsidization, especially those other than lighting, are mostly received by the relatively better-off households in the rural areas. (Although, indirectly, by increasing demand and the load factor, subsidization may lead to benefits for other sections of the community as well.) It is more likely, however, that the availability and reliability are more important, particularly from new consumers. This is because electricity is only part of the total costs of using electric power, whether for households or for business, and because energy costs are only a fraction of total costs.

An illustration of this is provided by the use of electricity in irrigation. The costs of electricity are minor compared with the fact that capital costs for electric motors are generally higher than those for diesel motors, while maintenance and nonfuel operating costs are higher for diesel pumps. So subsidizing electricity rates would probably have a minor effect unless the costs of connection and pumps are themselves subsidized, or credit provided on favourable terms. Some industries may also locate in rural areas if electricity is cheaper. This applies especially to industries such as cement and pulp and paperwood, with high energy content and input requirements which can be met in rural areas. Even for these, reliability of supply is probably as important as the price. An unexpected finding concerning small-scale industries is that their fuel costs are much higher than for others; thus, changes in the price of electricity would probably have some favourable effects on their operations (see SIETI, 1978). This may be expected to lead to increases in rural employment. It is true that some fragmentary evidence suggests that the benefits may not be as large as conventionally assumed, especially in the short run (SIETI, 1978). However, this is more due to the fact that regardless of the subsidization of electricity, the producers may not sufficiently increase their operations because of various other factors. One main issue which has been mentioned concerns the problem of finding markets. This calls for various other provisions, such as cheap transportation, which would strengthen rather than weaken the case for the provision of cheaper electricity to these industries.

Pricing of traditional fuels

Most rural communities are largely closed systems with respect to energy. It is estimated that as much as 70 to 90 percent of the energy used in several Asian countries is still obtained from local sources, with firewood, cattle dung, and crop residues contributing the bulk, and human work and animal work contributing the balance.[15] Household activities, in particular cooking, account for the largest amount of traditional fuel demands. Although these fuels are not used exclusively by the poor, higher income groups generally rely more on kerosene or electricity. A host of cottage industries and small-scale enterprises also use traditional fuels, in particular firewood.[16]

Since the price of these fuels is determined by the market, government policy by pricing substitutes may be expected to have some effects on their consumption. One of the major issues here is the serious shortfall in the supply of firewood, the rise in its price, and the deforestation problem. Although the situation varies from country to country, and even from region to region, there is little doubt that the deforestation problem is extremely serious. On average, forests have been diminishing by more than 1 percent per year, while population has been increasing by more than 2 percent (Barnes *et al.*, 1982). Deforestation is a major element in soil erosion, which has been increasing at a rapid pace in many countries. This destroys the soil, with serious implications for agricultural productivity. Firewood scarcity also affects agricultural productivity by forcing farmers, especially on small farms, to stop using animal and crop residues as fertilizers and soil conditioners, and to use them instead as fuels.

Could the rise in the price of firewood with its serious implications for poor people's budgets be influenced by government pricing policy? The most obvious method may seem to be subsidized kerosene and electricity. But, as noted earlier, it is not clear that this would have a significant effect on fuelwood price. This is because the markets for firewood and commercial fuels are highly segmented and the elasticity of substitution between fuels is probably quite low. A related aspect is that there are significant costs involved in buying equipment to use commercial fuels. This, as well as the social set-up and the living conditions of the poorest people in rural areas, is likely to preclude any significant switch to commercial fuels.

In the long run, the solution may appear to lie with increasing the availability of firewood by public reforestation programmes. With increases in prices, it has also become commercially viable for farmers to increase both the supply from the existing stocks and the land given over to woodlands. There is little likelihood, however, that this will be undertaken on a sufficient scale. In the short run, it may appear desirable to set maximum prices for fuelwood. However, apart from possible adverse effects on production, this is likely to be an unrealistic solution. There is the immediate problem of deciding what the maximum prices should be. The quality, type, and heat content vary enormously. Further, since the markets are highly localized, the maximum prices may be simply unenforceable. Nevertheless, if the situation is as desperate as some commentators suggest, it may be appropriate for a state agency to intervene directly by buying the firewood and selling it to the poorest people at subsidized rates. The size of the total subsidy is likely to be very small, but it could make a considerable difference to the welfare of the poorest people.

LIMITATIONS OF ENERGY PRICING

There is a long-standing debate in development economics on whether consumers and producers in developing countries respond to price signals in the manner suggested by traditional theory. In general, the evidence seems to indicate that they do, but that the price signals themselves are distorted. In the case of energy the distortions may appear to be imparted by government's desire to pursue various conflicting objectives; but, in reality, it is not even clear that the agents can respond to the signals. We have noted various instances of this: an increase in the price of kerosene, for example, may not necessarily lead to consumers switching to alternative, cheaper fuels, or an increase in the price of diesel need not result in industrial firms economizing in any significant way. The main reasons are, of course, that it requires sufficient information about the alternatives, and it requires their economic availability. In the absence of these, there may be very little that price manipulations can achieve on their own.

This suggests that a part of the energy pricing strategy must be to provide adequate information on alternative fuels, on alternative technologies, and on the energy constraints facing the economy. If prices were playing their role, the efficiency with which energy is utilized, whether in households or in the productive sectors of the economy, would have improved considerably over the past decade. Considerable evidence indicates that this has not been the case. Of course, information itself and the provision of alternatives may not be sufficient. Sometimes changes may be required in the social set-up and with work practices. This must also be taken into account.

One could go further and suggest that, whether or not these other preconditions are satisfied, it may become necessary to rely on nonpricing mechanisms to achieve various objectives. Mechanisms such as rationing of fuels are difficult to administer and introduce various distortions. Leakages are inevitable, but despite these such mechanisms may be necessary. Other mechanisms could be used in conjunction with the pricing strategy. These include direct regulations (such as maximum speeds for autos and trucks), strict adherence to fuel efficiency criteria for vehicles at the time of giving operating permissions, and establishment of efficiency standards. Because of the uncertainty regarding the efficacy of the pricing strategy, it may be appropriate to use, in addition, the nonpricing mechanism for both conserving energy and for influencing the fuel mix.

CONCLUDING COMMENTS

This chapter has emphasized that, in general, since the economies of the developing countries are far from being free from distortions, any pricing policy should be more concerned with being internally consistent, while taking into account explicitly the socio-economic goals being pursued and the major constraints on them. This means that the classical efficiency criteria, while not irrelevant to the analysis, may have to be given rather less importance. This also highlights the need for undertaking detailed empirical analysis. A great deal of research has been done into the energy sector in individual countries. Still very little is known, however, about the response of different agents to energy price changes. Research in this area must be an overriding priority.

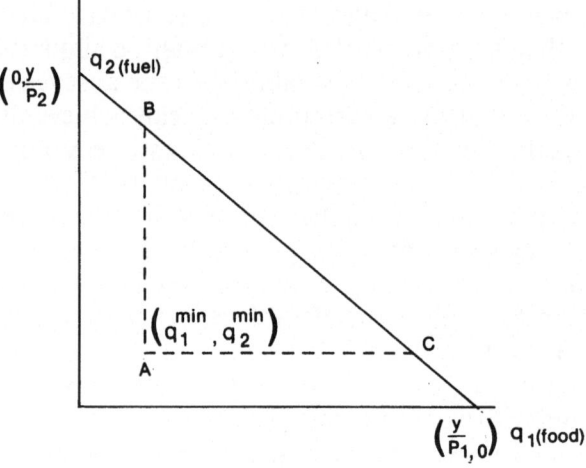

Figure 2.1

NOTES

[1]See, for example, Kumar and Panic (1983).

[2]This is the familiar second-best problem of Lipsey and Lancaster (1956).

[3]The discussion was originated by Hotelling (1938). For a more recent formulation, see Baumol and Bradford (1970).

[4]For a summary of the evidence, see, for example, Fields (1980).

[5]A speech by Robert MacNamara, then President of the World Bank, set the scene in 1973. For an excellent review of the issues, see Paukert *et al.* (1981).

[6]Most of the literature examines distribution of nominal income, implicitly assuming that changes in the price level affect different income households in the same way. This assumption is invalid, since it is very likely that inflation is regressive in its effects. See Williams, ed. (1977), especially Chapter 10 by Pond.

[7]A focus on the real income of the poorest households can also be obtained, of course, by giving them much greater weight in certain measures of distribution.

[8]It may be thought that progressive income taxes would be more appropriate. However, because of the generally low level of incomes, income taxes are levied much less frequently. See Toye (1978), Chapter 1.

[9]The analysis was undertaken for the Islamic Republic of Iran, Malaysia, the Philippines and the Republic of Korea, and the results apply to the last three of these countries. The elasticity of employment with respect to redistribution of income was found to be around 0.3 to 0.5; that is, a progressive redistribution of 10 percent of total income would raise employment by 3 to 5 percent. The redistribution is equivalent to a reduction in the Gini coefficient from 0.5 to 0.4.

[10]There is a burgeoning literature in this area. For a summary, see Mork (1981).

[11]In the short run, reduction in profitability may also have some adverse consequences for investment. See Kumar (1984).

[12]See also Berndt and Wood (1975 and 1979), Rasche and Tatom (1977), and Hudson and Jorgenson (1978).

[13]In the case of a country exporting liquid fuels, there may be substantial increases in foreign exchange earnings.

[14]As Sebastian (1979) has emphasized, the problem often lies in the choice of the product being produced and the associated investment, rather than in the type of technology. In many cases, there may not be much freedom for technological choice (pp. 67–72).

[15]See Hughart (1979) for detailed evidence.

[16]Traditional fuels are also used in the modern sector; for example, charcoal is used for steel-making in the Philippines.

REFERENCES

Ahluwalia, M. S. and N. G. Carter (1979). "Growth and poverty in developing countries" in H. Chenery, *Structural Change and Development Policy.* New York: Oxford University Press.

Atkinson, A. B. (1975). *The Economics of Inequality.* Oxford, UK: Clarendon Press.

Barnes, D. C., J. C. Allen, and W. Ramsay (1982). *Social Forestry in Developing Countries.* Discussion Paper D-73F. Washington, D.C.: Resources for the Future.

Baumol, W. J. and D. F. Bradford (1970). "Optimal departments from marginal cost pricing." *American Economic Review* (June).

Behrens, A. (1984). "Energy and output implications of income redistribution in Brazil." *Energy Economics* (April).

Berndt, E. R. and D. O. Wood (1975). "Technology, prices, and the derived demand for energy." *The Review of Economics and Statistics* (August).

Berndt, E.R. and D.O. Wood (1979). "Engineering and econometric interpretations of energy-capital complementarity." *American Economic Review* (June).

Berry, A. R. and W. R. Cline (1979). *Agrarian Structure and Productivity in Developing Countries*. Baltimore: The Johns Hopkins University Press.

Bhalla, A. S., ed. (1979). *Towards Global Action for Appropriate Technology*. Oxford, UK: Pergamon Press.

Bhalla, A. S., ed. (1981). *Technology and Employment in Industry*. Geneva: International Labour Organisation.

Cecelski, E. and S. Glatt (1982). *The Role of Rural Electrification in Development*. Discussion Paper D-73E. Washington, D.C.: Resources for the Future.

Chenery, H. B. *et al.* (1974). *Redistribution with Growth*. New York: Oxford University Press.

Cline, W. R. (1977). "Policy instruments for rural income distribution," in C. Frank and R. Webb (eds.), *Income Distribution and Growth in the Less Developed Countries*. Washington, D.C.: Brookings Institution.

Dasgupta, P. and G. Heal (1979). *Economic Theory and Exhaustible Resources*. Cambridge, UK: James Nisbet and Cambridge University Press.

Deaton, A. and J. Muellbauer (1980). *Economics and Consumer Behaviour*. Cambridge, UK: Cambridge University Press.

Desai, A. (1981). *Interfuel Substitution in the Indian Economy*. Discussion Paper D-73B. Washington, D.C.: Resources for the Future.

Dunkerley, J., C. Knapp, and S. Glatt (1981). *Factors Affecting the Composition of Energy Use in Developing Countries*. Discussion Paper D-73C. Washington, D.C.: Resources for the Future.

Eckholm, E. P. (1980). "The other energy crisis: firewood." in V. Smil and W. E. Knowland, (eds.), *Energy in the Developing World*. Oxford, UK: Oxford University Press.

Fallen-Bailey, D. G. and T. A. Byer (1979). *Energy Options and Policy Issues in Developing Countries*. Staff working Paper No. 350. Washington, D.C.: The World Bank.

Fields, G. S. (1980). *Poverty, Inequality and Development*. Cambridge, UK: Cambridge University Press.

Hotelling, H. (1938). "The general welfare in relation to problems of taxation and of railway and utility rates." *Econometrica* (July).

Hudson, E. A., and D. W. Jorgenson (1978). "Energy prices and the U.S. economy, 1972–1976." *Natural Resources Journal* (October).

Hughart, D. (1979). *Prospects for Traditional and Non-Conventional Energy Sources in Developing Countries*. Staff Working Paper No. 346. Washington, D.C.: The World Bank.

Hughes, G. A. (1983a). The Impact of Fuel Taxes in Thailand. Cambridge, UK: Faculty of Economics and Politics (July). Mimeo.

Hughes, G.A. (1983b). The Impact of Fuel Taxes in Tunisia. Cambridge, UK: Faculty of Economics and Politics (August). Mimeo.

Jankowski, J. E. (1981). *Industrial Energy Demand and Conservation in Developing Countries*. Discussion Paper D-73A. Washington, D.C.: Resources for the Future.

Jequier, N. (1979). "Appropriate technology: some criteria," in A. S. Bhalla (ed.), *Towards Global Action for Appropriate Technology*. Oxford, UK: Pergamon Press.

Kumar, M. S. (1983). The Financing of Energy Imports in Developing Countries. Cambridge: Department of Applied Economics. Mimeo.

Kumar, M.S. (1984). *Growth, Acquisition and Investment*. Cambridge, UK: Cambridge University Press.

Kumar, M. S. and M. Panić (1983). *International Interdependence and the Debt Problem*. Paper presented at 7th World Congress of the International Economic Association held at Madrid.

Lipsey, R.G. and K. Lancaster (1956). 'The general theory of second best.' *Review of Economic Studies*, Vol. 24, No. 63, pp. 11-32.

Mork, K. A., ed. (1981). *Energy Prices, Inflation and Economic Activity*. Cambridge, Mass.: Ballinger Press.

Munasinghe, M. (1980). "An integrated framework for energy pricing in developing countries." *Energy Journal* (July).

Newbery, D. M. G. (1981). "The Taxation of Oil Consumption." Report commissioned by the Policy Review Unit, British Petroleum. London: 20 July 1981. Manuscript.

OECD (1982). "Why do substitution elasticities differ?" Discussion prepared for the *Energy Seminar*, New York, December 1982. Paris: OECD.

Ostry, S., J. Llewellyn, and L. Samuelson (1982). "The cost of OPEC II." *The OECD Observer* (March), pp. 37–39.

Paukert, F., J. Skolka, and J. Maton (1981). *Income Distribution, Structure of Economy and*

Employment. London, UK: Croom Helm.

Ramesh, J. and C. Weiss, Jr., eds. (1979). *Mobilizing Technology for World Development*. New York: Praeger Publishers.

Rasche, R. and J. Tatom (1977). "Energy resources and potential GNP." *Federal Reserve Bank of St. Louis Review* (June).

Saicheua, S. (1984). *Input Substitution in the Manufacturing Sector of Thailand: Implications for Energy Policy*. Resource Systems Institute Working Paper WP-84-9. Honolulu: The East-West Center.

Satsangi, P. S. and V. Gautam (1983). *Management of Rural Energy Systems*. New Delhi, India: Galgotia Publications.

Sebastian, L. (1979). "Appropriate technology in developing countries: some political and economic considerations." in J. Ramesh and C. Weiss (eds.), *Mobilizing Technology for World Development*. New York: Praeger Publishers.

Shell (1983). Energy Demand Matrices: East Asia. London, UK: Shell International PLC. Mimeo.

SIETI (Small Industry Extension Training Institute) (1978). *Prospects for Modernising Rural Artisan Trades and Decentralised Small Industries*. Hyderabad, India: Yousufguda.

Smil, V. and W. E. Knowland, eds. (1980). *Energy in the Developing World*. Oxford, UK: Oxford University Press.

Solow, R. M. (1974). "The economics of resources or the resources of economics." *American Economic Review* (May).

Squire, L. (1981). *Employment Policy in Developing Countries*. New York: Oxford University Press.

Stewart, F. (1981). "Manufacture of cement blocks in Kenya." in A.S. Bhalla (ed.), *Technology and Employment in Industry*. Geneva: International Labour Organisation.

Stiglitz, J. E. (1975). "The efficiency of market prices in long run allocation in the oil industry." in G. M. Brannon, *Studies in Energy Tax Policy*. Cambridge, Mass.: Ballinger Publishing Company.

Toye, J. F. J. (1978). *Taxation and Economic Development*. London, UK: Frank Cass.

Turvey, R. and D. Anderson (1977). *Electricity Economics*. Baltimore: The Johns Hopkins University Press.

United Nations (1983). *Yearbook of World Energy Statistics 1981*. New York: United Nations Publication, Sales No. E/F. 82 XVII.16.

Visaria, P. (1980). *Poverty and Unemployment in India: An Analysis of Recent Evidence*. Staff Working Paper No. 417. Washington, D.C.: The World Bank (October).

Williams, F., ed. (1977). *Why the Poor Pay More*. London, UK: Macmillan.

Wiseman, J. (1957). "The theory of public utility price — An empty box." *Oxford Economic Papers* (February).

World Bank (1980). *Energy in the Developing Countries*. Washington, D.C.: The World Bank (August).

World Bank (1983). *World Development Report*. New York: Oxford University Press.

Chapter 3

EFFICIENCY AND EQUITY CRITERIA IN ENERGY PRICING WITH PRACTICAL APPLICATIONS TO DEVELOPING COUNTRIES IN ASIA

David M. G. Newbery

This chapter shows how to apply the general principles of energy pricing to the practical problem of setting the prices of particular fuels in developing countries, specifically those in Asia. The emphasis is on setting the prices to energy *users,* rather than designing the price and tax system which will induce the right level of exploration, extraction, and supply of energy.

The structure of the chapter is as follows: The first section argues that, under certain conditions, producer prices should be determined by efficiency considerations, while equity considerations are taken into account in setting consumer prices. The important differences between developed and developing countries are then identified, and the case for identifying the efficient price as the short-run marginal cost is argued. The second section discusses how to set the producer price of energy under a variety of circumstances. The third section examines the problem of separating consumer and producer prices of energy, which is mainly a problem for pricing kerosene and diesel. The fourth section discusses the problem of setting the price of rural electricity where equity and efficiency considerations are most in conflict. The last section draws conclusions.

PRODUCER AND CONSUMER PRICES OF ENERGY

Energy users can be either producers of other goods (such as steel mills, trucking companies, farmers) or final consumers (households using fuel for cooking, lighting, private transport). The first key distinction to make in the design of energy pricing policy (or, indeed, any pricing policy) is the distinction between *producer prices* and *consumer prices*. Producer prices are the prices facing producers who buy energy to produce other goods, while consumer prices are the prices facing final consumers. Producer prices are also the prices facing producers who sell energy, such as refineries, coal mines, and power stations, and, for many purposes, it is not necessary to distinguish between producers who buy and those who sell energy. The emphasis in this chapter is, however, on those who buy energy.

The importance of the distinction between producer and consumer prices is this. In a competitive economy in which profits (and rents) are either negligible or adequately taxed[1] and in which the government can effectively separate consumer and producer prices, producer prices should be set on efficiency grounds, and equity considerations will only be relevant for setting consumer prices. The difference between producer and consumer prices is then an indirect tax, and the design of energy prices divides into two parts: setting producer prices at the efficient level, and choosing the right set of taxes on energy to give the right set of consumer prices.

The distinctive differences between energy pricing policy in developed and developing countries can now be stated. In developed countries with a potent tax system, there is a presumption that equity objectives can be best met through the direct tax system, leaving little role for indirect taxes on energy to redistribute income.[2] In developing countries, direct taxes play a relatively minor role, and hence equity considerations are of considerable importance in the design of indirect taxes in general, and energy taxes (and subsidies) in particular. Second, while a comprehensive value-added tax system is an effective way of separating producer and consumer prices, and leaving producer prices undistorted and efficient, and while such a system is feasible in many developed countries, it is not so obviously feasible in developing countries, with two important consequences. First, where it is not possible to separate consumer and producer prices for energy, it is no longer possible to separate equity and efficiency criteria. Second, even where it is possible to separate the consumer and producer prices of particular fuels, if other inputs into production are taxed, it is no longer so straightforward to calculate the efficient price of the fuel with which to confront the producer.

Finally, developing countries typically experience more market distortions than developed countries, further complicating the calculation of efficient prices. One of the most important distortions in many developing countries lies in the market for foreign exchange which is particularly important for the pricing of imported and domestic energy sources, notably for petroleum products as compared with indigenous gas.

Of course, there are market failures which are common to both developing countries and developed countries and which are also important for energy pricing policy. One of the most important is the use of the pricing system to induce efficient production in state-owned or regulated enterprises — a pervasive problem in the energy sector. Since this is a contentious issue on which the various contributors to this book are not agreed, it may be useful to give a concrete example here, that of setting the producer price of electricity.

Efficiency and the pricing of electricity

The *efficient* price of electricity is the *short-run* marginal (social) cost of producing the electricity, and, if production cannot be increased, it is the price at which demand is equated to the given supply. It is important to stress that this is in general not equal to the long-run marginal cost, and that when the two differ, the long-run marginal cost is not the efficient price. There is a long history of confusion on this point, which would be tedious to relate, mostly arising out of a failure to distinguish various aspects of the problem of optimizing electricity supply. Several points, however, can be made fairly briefly.

First, the short-run marginal cost is well defined by the existing stock of equipment and options open in the short run (and, indeed, is typically carefully calculated in determining the merit order of power stations). The long-run marginal cost is not as well defined, since it is a forward-looking concept based on expectations as to the best choice of investment to expand the system. Second, if demand were constant throughout the year, and if investment were optimally undertaken with no indivisibilities, then the two marginal costs would be identical. Given fluctuations in demand, it remains true that the average short-run marginal cost would be equal to the long-run marginal cost, while with lumpiness in investment, under certain circumstances, on average, short-run marginal cost would equal the average long-run marginal cost. The long-run marginal cost is best seen as a shorthand for an investment rule: invest when long-run marginal cost is below average short-run marginal cost. This is equivalent to the correct rule of investing when selling the extra electricity at short-run marginal cost which, if above the long-run marginal cost, will yield a positive net present value on the marginal cost of expansion. From this it follows that if investment decisions are on average correct, then pricing at short-run marginal cost *will* cover (marginal) interest costs, and, assuming constant returns to scale in investment, will cover total costs and earn the efficient rate of return on investment. (Economies of scale may be important in developing countries and raise further problems discussed below.)

Third, proponents of long-run marginal cost pricing concede the need for "promotional" pricing in the presence of excess capacity, and recognize the need to ration limited supply by raising prices in the face of excess demand. It would be simpler to abandon the incorrect principle of long-run marginal cost pricing and replace it by the correct short-run marginal cost which deals with both cases automatically.

Finally, the main defense of long-run marginal cost pricing is that it gives correct signals to consumers for investment and avoids the instability of short-run marginal cost pricing. Both objections to short-run marginal cost can be met by offering contracts of varying length, during which an agreed quantity of electricity is sold at an agreed stable price. Variations in consumption above or below this contracted amount would be priced at the spot price, or the short-run marginal cost. Finally, the spot price could be quoted as a discount to or premium above the long-run average price, which would be the long-run marginal cost. Consumers would then have a planning price for investment decisions and a decision price for short-run consumption decisions.

This concept of the efficient price is fine for electricity consumers but may not be sufficient to ensure efficiency in the supply of electricity. If, as has been proposed for large developed countries like the United States of America or the United Kingdom of Great Britain and Northern Ireland, electricity were produced in a large number of independent competitive generating stations, selling to consumers through a common carrier national grid, then prices would naturally be equal to short-run marginal cost, and competition would ensure cost minimization in the production and optimal investment decisions. This is impractical in developing countries where individual units are typically large relative to the market served, grids are often small and poorly integrated, and hence competition unrealistic. How, then, to encourage power companies to minimize costs, supply at the right level of reliability, and take sensible, timely investment decisions? The danger with short-run marginal cost pricing is that, if underwritten by subsidies when short-run marginal cost is below average cost, it provides an incentive to overinvest and underprice. If, on the other hand, the power company must finance its own investment, then it *may* underprice (if the rate of growth of demand exceeds the rate of interest) and will often underinvest.

The problem is best seen as a standard principal-agent problem: How best to ensure that the agent (the power company) performs satisfactorily for the principal (the government), given that the agent has detailed knowledge not readily available to the principal? If the principal intervenes too much, then this special knowledge will be inefficiently used, while too little control means that the agent can exploit the principal. Pricing is then part of the incentive system devised by the principal to encourage efficiency by the agent. When associated with limits on investment and borrowing, it may involve balancing the advantages of lower costs (pursued in order to generate funds for investment) against the inefficiencies of incorrect pricing (average cost pricing compared to short-run marginal cost pricing). Much of the discussion elsewhere in this book about the objective of *financial viability* comes under this heading of finding a satisfactory solution to the principal-agent problem in essentially noncompetitive markets.

SETTING THE PRODUCER PRICE OF ENERGY

Provided producer prices can be kept distinct from consumer prices, the principle to be followed is that producers should face efficient prices. This principle also applies to energy producers provided that the rents (the revenues derived from the difference between this price and the costs of extraction and/or production) can be taxed at a higher rate. This immediately raises two questions: What is the efficient price, and can rents be satisfactorily taxed in practice? The problem of determining the efficient price is a general one for all energy-pricing decisions and will be discussed below and elsewhere in this book, but several points can be noted immediately. The simplest case would be the enclave development of a fully internationally traded good (or "traded good" for brevity), such as oil or export liquefied natural gas. Provided the developer were free to purchase all inputs at world prices, the efficiency price would just be the export (or import) parity price of energy, which would be well defined and readily observable. At the other extreme, nontraded energy (such as hydroelectricity or domestic gas in countries with insufficient gas to justify LNG) often presents considerably greater problems in computing the efficiency price. Where the energy displaces imported energy at the margin, then the cost so saved measures the efficiency price. If hydro displaces oil-fired power generation or if gas displaces residual fuel oil in its marginal use as an underboiler fuel, then their prices are well defined by the relevant opportunity cost. If a large hydro site is to be developed far from the grid for use in, say, aluminium smelting (as in Papua New Guinea), then its efficiency price must be defined by its value in use: What is the highest price which the aluminium smelter can pay for the electricity and still earn a normal return? (In such cases, it may be logical to treat the smelter and the attendant power supply as a vertically integrated concern and tax the rent of the whole concern.)

The feasibility of rent taxes depends critically on the observability of the costs of inputs and the value of outputs and is extensively examined in the recent book by Garnaut and Ross (1983). (They also discuss the experience of rent taxation in Indonesia and Papua New Guinea.) Where large companies are involved, then the accounting base is well defined, but the main problem is likely to be transfer pricing, either for the output, if it is semi-processed and not priced on competitive markets, or for specific inputs (specialized rigs or machines). One solution is to prefer independent companies, but this may not be feasible.

Noncommercial energy is unlikely to be produced by companies with good accounting practices, but for most of these energy sources, rent is negligible, since they are usually renewable resources and hence like standard produced goods. Pricing some of the inputs correctly may be a problem, especially for fuels gathered from common land.

The efficient price of energy

If the government is successful in choosing taxes and tariffs so that producers face efficient prices for nonenergy inputs, then the efficient price
of any given fuel is, in theory, easy to calculate. For *traded* fuels (oil products pre-eminently), the efficient price is derived from the border price.
If the country imports diesel and exports gasoline, then the efficient prices
at the dockside are the c.i.f. price for diesel and the f.o.b. price for gasoline.
For nontraded fuels (gas, electricity) the principles are straightforward,
but their application is more complicated and is discussed elsewhere in
this book. The reason is that the marginal cost must be calculated at efficiency, or shadow prices, and the distinction between marginal and average
cost kept clear, even though most accounts only contain figures for average
costs. Thus in the case of electricity, the first question to ask is: What
is the cost of marginal (i.e., highest cost) supplies? Although coal may
be used for a large fraction of total production (as in India), if oil (or worse,
diesel) is used for marginal base load generation, then oil prices will determine the efficiency or shadow price of electricity. Coal is also quite difficult to shadow price, unless it is traded. Its shadow price is also very location
specific, as transport and handling costs are high.

Although it is easy to calculate the efficiency prices of petroleum products from the c.i.f. price (or, in cases where a local refinery exports the
product, the f.o.b. price), it is important to recognize that governments
frequently distort ex-refinery prices, causing cross-subsidization with different products selling at prices different from their opportunity costs (or
world prices). Table 3.1 illustrates this for India. When the world oil price
rose after 1973, the price of Indian crude remained pegged at the 1973
import parity price until 1976 and at Rs 45 per barrel after that. There
was thus a massive cross-subsidization of imported oil by local crude
reflected in the negative refiners' margins shown in Table 3.1. The effective tax on products should thus be considered as the sum of the tax and
refiner's margin (especially as all but one small refinery were state owned),
and, while taxes on this basis remained positive, they clearly fell
dramatically in *ad valorem* terms, as Table 3.2 shows. Diesel became
relatively cheaper than the still heavily taxed gasoline, while furnace oil
became relatively more expensive than diesel. Kerosene at market prices
was cheaper than diesel, though more expensive at border prices.

Implications for energy prices of severe revenue scarcity

Some readers will question whether the principle of confronting producers
with efficient (i.e., untaxed) prices will hold if the government is desperately
seeking ways of increasing tax and other revenue. If electricity is sold
at short-run marginal cost, then how will the power sector's investment
needs be met? The answer is that, as far as possible, it is preferable to
levy taxes on consumers, not producers (except for taxes on rents and
pure profits). If the government is seeking further tax revenue, then it

should increase taxes on goods with a high income elasticity and low price elasticity.[3] Domestic electricity, gasoline, and the durable goods which use electricity or gasoline are potentially attractive candidates for raising taxes (i.e., raising the consumer price).

If revenue is very scarce, then the government will be constrained in the size of its investment programme and will have to ration the scarce investible resources by raising the rate of discount. This in turn will mean less investment in, e.g., power generation, hence less supply, and a higher market clearing price, which will generate higher revenues for the power company and alleviate the financial constraint. Put another way, the long-run marginal cost will increase because of the rise in interest rates, and the short-run marginal cost (and the price) will gradually increase as investment is delayed and demand grows. The most perverse solution, which is commonly observed, would be to restrict investment while not raising prices, leading to excess demand, rationing, and unreliability in the power supply. As Schramm shows in Chapter 4 of this book, this can be exceptionally costly.

Finally, a scarcity of government revenue will make the opportunity cost of such revenue higher and will reduce the desirability of subsidies (e.g., for kerosene).

The efficient price of energy in the presence of distortions

If, however, producers do not face efficient prices for nonenergy inputs, then in general they should not face border prices for energy. There are then two possibilities. The better solution is to correct the existing distortions facing producers, so that major energy users in particular face the correct nonenergy prices. In general this will involve reforming tariffs, quotas, exchange controls, etc.: in short, liberalizing the trade regime of the economy. An equivalent but less drastic solution would be to allow producers to recover duty on the purchases of imported inputs. Only if this strategy fails would it be desirable to adopt the second and inferior alternative, which is to set the price of energy in such a way as to offset the presumably irremediable inefficiencies elsewhere. Consider two examples. Suppose a country taxes rice (i.e., pays the farmers well below export parity) and that rice production requires energy inputs (for tubewells and, indirectly, for fertilizer). Then it may be desirable to subsidize energy purchased by farmers in order to bring the relative prices of inputs (energy) and output (rice) closer to the relative border prices. More generally, it is unlikely that the efficient price of energy sold to rice farmers should be the border price, though it will in general be quite a complex exercise to establish the right price for energy, since it will depend on the extent to which the prices of other inputs, notably labour, can be altered.

The second example illustrates some of the other difficulties involved in pricing in a distorted economy. Suppose steel production is protected by an import tariff, which cannot be altered. If the domestic steel producer is a monopolist, and if he sets the sales price of steel at the border price cum tariff level, thereby making large profits, it may be desirable to tax his fuel purchases to the point at which his profits fall to a normal level. If, on the other hand, there are several vigourously competitive domestic producers selling steel at above the border price, but below the border price cum tariff, then it may be desirable to subsidize energy purchases in order to lower the domestic price of steel to the border price level. In both cases it would be preferable to eliminate the tariff on steel, and, if it is argued to be essential to produce steel domestically, and assuming there are no other distortions, to subsidize steel production directly, rather than through a subsidy on inputs.

It should be clear that the second best approach, in which energy prices are adjusted to offset existing distortions, can be highly complex and will, in general, require different producers to face different prices for the same energy. While this may be possible for gas and electricity, which are not easily resaleable, it may be impossible for liquid and solid fuels. The preferable approach would therefore seem to involve setting producer prices of energy at their efficient levels and then dealing with resulting inefficiencies directly by adjusting other prices (i.e., eliminating other distortions). In some cases, this will require production to close down (notably for the production of fertilizers by inefficient or obsolete processes; Egypt, for example, has an ammonium plant based on the electrolysis of water) and may therefore require compensation to be paid to the plant owners (if, as for some fertilizer producers, the government gave a commitment to guarantee the price and sales levels of the output).

Second-best pricing

Is it possible to calculate the optimal producer price of energy in a distorted economy in which it is impractical to eliminate the major distortions? This question has been addressed in Dixit and Newbery (1985). They considered a simple general equilibrium model of Turkey in which the relevant distortions consisted of tariffs and producer taxes, and they showed that the optimum tax to place on oil was a weighted average of these taxes and tariffs. (The tax on oil was to be applied to the border price, or efficient price — also its shadow price — to bring the price of oil to producers up to the level which minimised aggregate inefficiency. In general, additional taxes aimed at consumers would be needed, but these were ignored.) The interesting finding was that, although the weights to apply to the existing taxes and tariffs had to sum to unity, they did not have to be positive nor individually less than one, and so the weighted average could be outside the range of existing taxes. This is an important finding, for it is often argued that since oil is often more heavily subsidized than any other commodity, it must be desirable to reduce the subsidy. In our model it can be desirable to subsidize oil to offset the excessive taxes on the production of oil-intensive goods.

We found that for Turkey this was not just a theoretical possibility, but apparently justified by the data; despite positive taxes on all other sectors, our best point estimate of the tax on oil was -12 percent, or a subsidy of 12 percent of the border price. However, this figure was extremely sensitive to a wide range of parameters in the model, most of which, such as the elasticities of substitution between capital and labour in each sector, are known with low accuracy. A realistic confidence interval for the tax on oil was estimated as -50 percent to 30 percent, which is far too wide to give one much confidence in advising on the correct price of oil alone. The conclusion to draw from this is that it is very difficult to set the price of energy in a highly distorted economy and that it is a rather perverse exercise to attempt to reduce these distortions by adjusting only the price of energy. The correct conclusion is to address the major distortions directly, and the less distorted the prices facing producers, the easier it will be to set the correct producer price for various fuels.

For the rest of this chapter, we shall therefore assume that it is desirable, as far as possible, to confront producers with efficient prices and deal with other distortions directly. There are, however, two further problems even if the other distortions facing producers can be dealt with. First, it may be impossible to charge consumers and producers different prices, in which case it will in general no longer be desirable to set the common price at the efficient level. Second, the efficient price for the same fuel in different uses may differ. The primary example would be diesel, which as a road fuel may be the selected method of charging for road use, while in other uses (tractors, generators, stationary motors, heating fuel, etc.) this argument would not apply. Both issues raise similar problems which we now address.

THE DIFFICULTY OF SEPARATING PRODUCER AND CONSUMER PRICES

Most fuels are consumed solely by either producers (coal, lignite, fuel oil) or consumers (gasoline), or readily sold at different prices to consumers and producers (gas, electricity). The only fuels which are sold to both consumers and producers and for which it is difficult to charge different consumer and producer prices are diesel and kerosene. Similar problems arise when it would be efficient to price discriminate between fuel for road transport use and the same, or similar, fuels (e.g., kerosene) used elsewhere. Again, as far as energy pricing is concerned, the only problematic fuels are diesel and kerosene.

The key issue is that automotive diesel fuel is a natural tax base in charging for road use (though one which must be supplemented by vehicle taxes and annual licence fees) and hence should be priced above border parity, while diesel used in industry and agriculture should be priced at border parity on efficiency grounds (though, as we have seen, it may be desirable to price diesel in agricultural use at below border parity if agricultural prices are below border parity). Finally, to further complicate matters, it has sometimes been argued that kerosene should be subsidized for two quite separate reasons. First, if kerosene is an inferior fuel (to electricity for lighting, for example) or an essential fuel (for cooking), then a tax on kerosene may be regressive, and a subsidy would improve the distribution of income and hence be attractive on equity grounds. Second, kerosene may be substitutable with woodfuel which may be underpriced as it is collected from communally accessible forests. The true social cost of collecting woodfuel may be well above the private cost, since the collectors do not have to pay for the cost of producing the trees, nor for the subsequent ecological degradation which may result from excessive deforestation. In order to confront consumers with the correct relative prices of woodfuel and kerosene, it may therefore be necessary to subsidize kerosene.

However, it is possible to substitute kerosene for diesel to some extent in automotive use (maybe up to 30 percent) and very easy to do so in heating uses, so the subsidization of kerosene in turn affects the price which should be set for diesel.[4] It is therefore interesting to consider the pricing of kerosene and diesel in some detail.

Setting the price of kerosene in Thailand

Hughes (1983) has examined the effects of changing the price of kerosene in Thailand using the 1975 Thai Input-Output Table (NESDB, 1980) and detailed household budget data from the Thailand Socio-Economic Survey of 1975–1976. (The methodology of such an impact study is set out in the annex to this chapter, and further illustrated there.) If the price of kerosene is increased by 50 percent and consumption patterns remain unchanged, then a household currently spending X baht per month will pay an extra percentage amount, T/X

$$\frac{T}{X} = 0.843 - 0.95 \, (X/10,000), \qquad R^2 = 0.05. \qquad (1)$$

$$(55.7) \qquad (24.1)$$

(Brackets give t-values, sample size 11,000. Hughes, 1983). The negative coefficient on expenditure implies that kerosene taxes are indeed regressive, but there are two points to notice. First, the low R^2 of 5 percent means that the main effect of a kerosene price change is uncorrelated with income. Second, the price impact will be extremely small, as kerosene expenditure accounts for only 2.4 percent of total expenditure on all petroleum products, which account for about 5 percent of gross national product (GNP). Annex Figure 3.1 demonstrates the relationship between the percentage cost increase caused by the kerosene price rise and income level. The regression line is drawn, together with a parallel line below which 90 percent of the 11,000 observations lie. The great bulk of the observations lie in the shaded area.

A very small fraction of the sample population would face cost increases of more than 2 percent, and more than 90 percent would face cost increases of less than 1.5 percent for what would be a very sizeable price rise for kerosene. In fact, in Thailand kerosene prices have not been subsidized, and, perhaps as a result, kerosene consumption is modest. In some other countries kerosene has been heavily subsidized, consumption levels are considerably higher, and the problem of raising kerosene prices might be more severe.

It is interesting to see if the adverse effects on equity of raising kerosene prices could be offset by subsidizing (or reducing the tax on) some alternative consumption good. What is required is some other good for which taxes are regressive, and food is the obvious such example, though if food is already subsidized (through the export tax on rice, for example), then the efficiency costs of increasing the subsidy may argue against that choice. Ahmad and Stern (1983) have developed a methodology for identifying the direction in which it would be desirable to change taxes, allowing for equity and efficiency considerations, and this methodology could, in principle, be used to identify the set of goods whose prices could be lowered to more than offset the effect of raising the price of kerosene.[5]

To this end we computed Engel curves for various goods consumed in Thailand to identify necessities. The following equation was estimated from the consumer budget survey data. Table 3.3 presents the results.

$$w = \alpha + \beta N + \gamma \log (X) \tag{2}$$

where w is the budget share,
       N is the household size,
       X is real expenditure.

If the parameter γ is negative, the commodity is a necessity, and subsidies (or reduced taxation) will improve the distribution of income. The third column of Table 3.3 gives the correlation between expenditure on the good identified and kerosene, and measures the ease with which the adverse distributional impact of raising kerosene prices can be offset by subsidies to that good. What comes over very clearly is that it would be hard to devise a neutral tax change which would leave everyone better off. Although it should be relatively simple to find a way of improving the distribution of income while raising the price of kerosene (and using the proceeds to reduce other taxes), this change would nevertheless have a fairly random effect — some households would gain while others would lose, despite having the same initial standard of living.

This analysis may be summarized as follows. There is a potential conflict between equity and efficiency in pricing kerosene, since efficiency requires a price as high as diesel, which, on efficient grounds, should be above the border price to recover some of the road-user costs. Kerosene is a necessity with a low-income elasticity, usually relatively more important to rural consumers, and hence an appealing choice of a good to subsidize on equity grounds. However, the expenditure share is typically small, so it is a relatively ineffective method of redistributing income. Moreover, its consumption is very poorly correlated with income, so it is a very imprecise method of directing income at the poor, and in absolute terms wealthier urban dwellers gain more from kerosene subsidies than poor rural consumers. Since the price elasticity appears to be quite high (perhaps unity), the inefficiency associated with large subsidies will be significant, quite apart from its effect on the adulteration of transport diesel fuel. Further, the magnitude of and redistribution that might be achieved by subsidizing kerosene is almost certainly negligible compared with the random shocks caused by the rates of inflation and exchange rate changes prevalent in countries which subsidize kerosene.

The taxation of transport fuels

The case for taxing transport fuels has two components — efficiency and equity. On equity grounds, it is clear that gasoline consumption in developing countries (and certainly in Thailand) is income elastic, and hence an obvious candidate for taxation. On efficiency grounds, vehicles incur costs on roads and other vehicles (both by making the roads rougher and hence more costly to other road users, and by congestion). There is thus an argument for charging vehicles for the use of roads, ideally an amount equal to the social costs caused by the vehicle. To the extent that these costs are related to distance driven, they can be recovered through fuel taxes. However, this is an inadequate tool by itself, as the damage done by vehicles rises as the fourth power of the axle loading, while fuel consumption per mile rises roughly linearly with gross vehicle mass. A given fuel tax per litre will thus undercharge heavily laden vehicles relative to lightly laden vehicles. The solution is to impose taxes on new vehicles and annual licence fees which reflect the damaging power of a typically laden vehicle. The tax on new vehicles has the effect of raising operating costs per annum on new vehicles relative to old, and since newer vehicles have higher utilization rates than old vehicles, this goes some way to making taxes reflect the different degrees of utilization. Alternatively, licence fees could vary with the age of the vehicle and this would reduce the financial constraints on the purchase of vehicles. Taxes on tires are potentially an attractive way of recovering road-use costs but have obvious drawbacks in terms of encouraging excessive wear and retreading, to the possible detriment of efficiency and safety.

The equity aspect is fortunately easy to separate from the efficiency aspect, since gasoline is primarily used in private cars, and diesel in commercial vehicles. To dissuade private car owners from choosing inappropriate diesel-engined cars, the annual licence fee for a diesel-powered private car should be set at a level mt above that for a gasoline-powered car, where t is the tax per litre on gasoline, and m is the number of litres consumption per annum at which diesel and gasoline versions are equally economic, costing both diesel and gasoline at their border or efficiency price. The main problem would probably be that people would buy diesel-powered commercial vehicles (pickups, vans, jeeps, etc.) for private use — an obvious inefficiency. This would limit the extent to which it would be sensible to tax gasoline and private cars relative to diesel and competitive commercial vehicles.

Congestion costs cannot easily be charged for through fuel taxes, and they require locationally specific charges — licences to drive in certain areas, licenses differentiated by address of owner, parking charges, and the like. Since congestion does not affect fuel pricing, we shall ignore it in this chapter.

The World Bank is currently developing a methodology for measuring the road-use costs incurred by various types of vehicles on various types of roads and is designing a suitable tax system to recover these costs; when this is ready (end of 1985), the principles of setting the efficient price of transport fuel should be better defined and operational.

EQUITY AND EFFICIENCY CONFLICTS IN ELECTRICITY PRICING

Electricity is typically an income elastic good and is often in short supply, with frequent power failures. There is thus a powerful case on equity and efficiency grounds for keeping the price high enough to ration demand by price rather than blackout. The more interesting conflicts arise when the power supply is experiencing economies of scale (i.e., capacity is built ahead of demand, or the hydro project has large indivisibilities). This is quite likely for isolated rural areas where the bulk of the costs will be in the equipment and infrastructure. The principle of charging for fixed costs by a connection charge then leads to decreasing average tariffs by use and will be regressive, while rural electricity is likely to be significantly more expensive on average than urban electricity, again likely to be regressive. One obvious solution is to make the fixed charge a function of installed capacity, which, if it is made progressive, will avoid the equity-efficiency conflict.

The pricing of rural electricity (and its provision) would merit a study on its own and would require extensive research. (For a useful survey of the present state of knowledge and suggestions for further research, see Cecelski and Glatt, 1982.) Several features stand out from this survey:

1. Rural consumption is low, less than one-fourth urban per capita levels. (In the Philippines, where rural electrification is considered very successful, 90 percent of connected households used less than 35 kWh/month — enough for two 100-watt bulbs for 4 hours/day [Cecelski and Glatt, 1982, p. 9].)
2. Often a very small proportion of households connect even when it is available (3.5 percent in rural Suryapet and 8 to 10 percent in

Karnataka, both in India [Cecelski and Glatt, 1982, p. 19]). Desai (1981, p. 44), however, cites a figure of 26 percent of households in electrified Indian villages in 1971.
3. Those connected have higher incomes than those who do not (between 2 to 4 times; Cecelski and Glatt, 1982, p. 21).
4. The main determinant of consumption is appliance ownership, and for durable appliances the cost of electricity is a small proportion of the total cost.

To gain some idea of the figures involved, Table 3.4 gives average and marginal costs for local and central electricity supply in El Salvador in 1975 (though it is not clear whether these costs are economic costs or include taxes and other distortions). It is clear that for low load factors and for centrally supplied electricity the excess of average over marginal cost is very large, posing considerable problems for tariff design. It is not, therefore, surprising to find that prices charged for electricity in rural areas vary greatly, from US$0.02/kWh in Nicaragua to US$0.16/kWh in Mauritania (Cecelski and Glatt, 1982, pp. 50–51), nor that subsidies are common, especially in early years. Thus the Rural Electrification Corporation of India expects negative returns on projects in "ordinary advanced" areas up to the sixth year and 3.5 percent returns by the end of the fifteenth year (Sengupta, 1979, p. 2).

It is also interesting to note that one of the main uses of electricity in rural India is for irrigation pumps, for which the alternative is a diesel pump set. The only cost-benefit comparison to use shadow prices finds that diesel pump sets are cheaper in Bihar than electricity, though at market prices the converse is true (except at discount rates of 15 percent or more) (Bhatia, 1979). The reason is that diesel is taxed relative to its efficiency price, while electricity is subsidized, as Bhatia discusses in Chapter 5 of this book. Since electricity sold for power use is typically priced differently from domestic electricity, its pricing can be guided by efficiency considerations, of which the main one is the price of diesel.

The main issue for rural electrification is probably deciding whether and when to electrify a village (and how). Once electricity is available, there is a case for charging above marginal cost for domestic consumption, as it is an income elastic good in relatively inelastic demand. There is also a case for an annual connection charge related to installed capacity, if that is feasible.

The main inequities likely to prove hard to deal with without high efficiency costs are the higher costs of electricity in rural areas or, worse, its nonavailability.

CONCLUSIONS

Conceptually the simplest fuels to price are those sold solely to producers in the formal sector (coal, fuel oil, lignite) or those for which price discrimination is feasible (gas and electricity). In such cases the ideal is to set producer prices at their efficient level. If other inputs and/or outputs are sold at distorted prices, and the only feasible instrument available to offset these distortions is the price of particular fuels, then there is an argument for adjusting these fuel prices, though it is hard to think of cases for which this is plausible. In other cases it is preferable to tackle the main source of distortion directly.

Where consumers can be confronted with a price different from the producer price (electricity, gas, gasoline), then equity considerations become relevant, and for these three fuels almost certainly argue for indirect taxes at above average rates to the extent that these fuels are income elastic and (moderately) price inelastic. The most problematic fuel to price is rural electricity, for which equity and efficiency principles give radically differing prices.

This leaves two fuels which raise special problems — kerosene and diesel. Since they are close substitutes in consumption, it is hard to price them very differently, and it is hard to separate the consumer and producer prices. On equity grounds, kerosene should probably be subsidized, while diesel should be (modestly) taxed on efficiency grounds to recover some fraction of road-use costs. If a government attached high priority to the distributional criterion, then the efficiency costs of underpricing diesel could probably be largely offset by increases in licence fees, made fuel- and capacity-specific to discourage inefficient substitution of large for small engines, or diesel for gasoline. If, on the other hand, it was felt important to move towards a more efficient set of producer prices (as part of a general tax reform, for example) then the adverse distributional effect of raising kerosene prices would be small and, on average, could be offset by the other tax changes. Subsidizing best practice kerosene-using equipment (lamps, stoves) might allow the losers to be more directly compensated.

As with all tax and price reforms, what is desirable depends sensitively on the range of possible reforms which can be simultaneously considered. The techniques are now available for identifying desirable reforms of the energy pricing structure (and of other prices and taxes) and are discussed in the annex in this chapter, in Ahmad and Stern (1983), and, more systematically, in Newbery and Stern (1985).

Table 3.1   Retail oil product prices in Calcutta, 1979 (Rs./kl)

	Foreign posted price	Taxes	Refiner's margin	Distribution costs	Total
Gasoline	1,009	3,095	−151	79	4 032
Kerosene	1,052	554	−315	89	1,380
High-speed diesel	989	1,338	−917	72	1,482
Furnace oil	650	146	184	51	1,031

Source: Desai (1981), p.54.

Table 3.2 Ratio of market price to efficiency price,[a] Calcutta

	1973	1975	1979
Gasoline	6.27	3.73	3.70
Kerosene	2.31	1.24	1.17
High-speed diesel	4.11	1.46	1.40
Furnace oil	2.48	1.30	1.47

Source: Desai (1981), p.54.
[a] Taken on the foreign price plus distribution cost.

Table 3.3 Necessities in Thailand

	γ	(t value)	Correlation with kerosene
Maize and cereals	−0.02	−42.4	
Charcoal and firewood	−0.01	−38.2	−0.07
Fish	−0.03	−39.8	−0.05
Canned food	−0.19	−39.0	−0.03
Milled rice	−0.14	−113.8	−0.05
Miscellaneous food	−0.01	−39.2	−0.05
Kerosene	−0.007	−50.0	

Table 3.4 Cost of rural electricity in El Salvador, 1975 (US dollars)

	Autogeneration		Grid			
Distance from grid	n.a.		4 km		29 km	
Load factor[a]	10%	25%	10%	25%	10%	25%
Variable cost ¢/kWh	6	6	1	1	1	1
Overhead cost ¢/kWh	15	6	17	6	39	16
Average cost ¢/kWh	21	12	18	7	40	17
Overhead fee $ per annum (for 400 kWh per annum)	60	24	68	24	156	64

Source: World Bank (1975).
[a] Ratio of average to peak consumption.

ANNEX 3.1
ESTIMATING THE IMPACT OF FUEL PRICE CHANGES IN THAILAND

If the market price of some good is very different from its efficiency price, then it is worth asking whether this market price is consistent with the policy objectives of the government, or whether these objectives indicate that the price should be changed. In a period when the world price of energy has changed dramatically relative to other prices, it is more than likely that energy taxes and pricing policy which may have been suitable before 1973 are no longer appropriate. The problem is that policy-makers face conflicting objectives (equity, efficiency, the need for government revenue, etc.) and, in a very distorted economy, find it difficult to assess the effect of any single price change on these various objectives. The World Bank has commissioned research designed to develop a method for assessing the effects of changing energy prices as part of a research project on the Pricing and Taxation of Road Transport Fuel (RPO 672–83). In this section, I shall briefly describe the method and the results of applying it to Thailand. Further studies of price impacts in Indonesia and Tunisia are currently under way.

The impact of fuel taxes on the distribution of income

The idea is simple in principle, but data- and computer-intensive in practice. We assume that there is no substitution effect as a result of the price changes and, as a result, the calculated impact is likely to overstate the true impact. All the evidence suggests that this effect is likely to be negligible for the size of price change likely to be politically feasible. The effects of any price change are traced through the input-output table to find the impact on the costs of producing the various goods in the economy. So far the technique is standard. Most authors of the technique assume, however, that increases in cost lead directly to corresponding increases in the prices of these goods, following a standard mark-up pricing rule. We argue, however, that in a small, open economy such as Thailand it is critical to distinguish between those goods that are priced on world markets, and those goods whose domestic prices can be set independently of world market prices. Consider two polar cases. Agricultural exports earn the world market price, and any increase in their cost of production or transport to the port leads to a fall in the incomes of the factors used in their production, and a fall in their farm-gate price (by an amount equal to the increase in transport cost to the port). Road transport services, on the other hand, are nontraded goods, produced under constant returns in competitive markets, and their prices will indeed be equal to cost and hence will increase as input costs rise. Charcoal is an intermediate case, for it is nontraded, but substitutable for traded kerosene. A 10 percent rise in kerosene price is assumed to lead to a 5 percent rise in charcoal price as consumers switch to charcoal and raise demand relative to supply.

Various tax changes can now be analysed, first calculating the tax rate which would be required to raise tax revenue by, say, 1 percent of gross domestic product (GDP) (so that alternative, equal yield tax changes can be compared), and then calculating the changes in prices of all goods resulting from these tax changes. Changes in factor incomes can be calculated from changes in gross output and input costs. These price and income changes can then be used to measure the change in real income of any household in the household budget survey (assuming no change in the quantities of each good purchased). The importance of using the whole sample (of 11,000 for Thailand) rather than a small number of "representative" household budgets at different income levels, or an estimated expenditure function, is that the likely diversity of impact of any tax change can be readily appreciated. Annex Figure 3.1 graphically illustrates the range of possible outcomes which results from a 50 percent increase in kerosene prices in Thailand. Some taxes (notably on intermediate goods) have a relatively more uniform impact on consumers than others. The most dramatic nonuniformity was found for an increase in export taxes, which lowers the domestic price of exportables (notably rice), benefiting urban consumers and having a relatively large adverse effect on large farmers. Small subsistence farmers are left relatively unaffected. Gasoline taxes have a fairly heterogeneous effect, as might be expected, and they are, as expected, quite progressive (indeed, the most progressive of taxes considered). A gasoline tax which yields 1 percent of GNP in extra tax revenue requires a sales tax at a rate of 46 percent and leads to a tax transfer of an amount T/X percent of expenditure, given by

$$T/X = \underset{(10.6)}{0.29} + \underset{(30.9)}{2.27} \quad (X/10000) - \underset{(5.4)}{0.03N}, \qquad R^2 = 0.08. \qquad (A.1)$$

(t values in brackets, $N =$ number of household members. Hughes, 1983, Table 8).

The heterogeneity of the impact can be examined by plotting a scatter diagram of T/X against X, as in Annex Figure 3.1 where the vast majority of the 10,000 observations lie in the shaded area shown, and by finding the lines parallel to the equation of T/X which encompass 80 percent of the observations, as shown for gasoline in Annex Figure 3.2.

If different tax changes are to be compared, then it is very useful to devise a summary statistic for the effect different equal yield taxes have on the distribution of income. Many such statistics are available, and there is a case for calculating several. Hughes presents four in his paper, all based on the Atkinson inequality index (see Sen, 1973). The conventional Atkinson indices of vertical inequality were computed with inequality aversion parameters $\epsilon = 1.0$ and 2.0 which correspond to relatively low and moderate aversion to inequality. The second pair of measures are King's (1983) index of overall inequality which extends the Atkinson measure to include a measure of the horizontal inequity generated by the tax change. While one may well have misgivings about either the relevance of horizontal equity for tax reform or the accuracy with which it is measured by the King index, it does, nevertheless, give a feel for the heterogeneity of the impact of the tax on households which are superficially similarly placed, but not similarly affected by the tax.

Annex Table 3.1 reproduces Table 8 from Hughes (1983) and gives the impact of imposing a number of tax changes in Thailand, all of which raise the same amount of government revenue — about 1 percent of final demand:

Reform number	Description
R1	A uniform sales on all petroleum products at 19.5 percent
R2	A sales tax on gasoline and aviation fuel at 45.7 percent
R3	A sales tax on all petroleum products other than gasoline and aviation fuel at 34.1 percent
R4	A uniform tariff on all nonfuel imports at 7.5 percent
R5	A uniform export tax on all exports at 8.9 percent
R6	A sales tax on all manufactured goods at 2.2 percent
R7	As RI but assuming a 1 percent fall in money wage rates and at rate of 18.5 percent
R8	As R6 but assuming a 1 percent fall in money wage rates and at rate of 2.1 percent

The last two alternatives assume that a rise in government revenue of 1 percent of GDP leads to a fall in money wages of 1 percent, reflecting one possible extreme case of the deflationary impact of the tax change. The others assume no change in money wages.

Section A of Annex Table 3.1 gives the coefficient of (X/10,000) and N in equations such as equation (A.1). The deviations reported in section B refer to the positions of the lines in Annex Figure 3.1 above and below the reference line. Sections C and D give the four indices of inequality mentioned above.

If we ignore issues of horizontal equity for the moment, the table reveals that gasoline taxes have the greatest impact on the distribution of income, and export taxes have the worst effect. There is, however, remarkably little to choose between the alternatives, and the main conclusion is that quite large tax increases on petroleum products have a modest but, on balance, beneficial impact on the distribution of income. If the government were to attach a moderate degree of importance to horizontal inequity, then at low levels of inequality aversion ($\epsilon = 1$) all reforms except the industrial sales tax make matters worse, though for higher inequality aversion ($\epsilon = 2$) fuel taxes improve equity. In turbulent times, when relative prices are changing, there is thus very little reason for not pursuing an efficient pricing policy for fuels and allowing them to rapidly adjust in line with fluctuations in international prices.

Impact of fuel taxes on inflation

Governments are not only reluctant to raise domestic fuel prices in response
to international price changes for equity reasons, but also because they
fear the inflationary consequences. Hughes calculated the impact of the
tax reforms listed earlier on the producer and consumer price indices in
Thailand, and they are reported in Annex Table 3.2.

The table makes clear that large price changes (20 percent increase in
the price of all fuels) have a small impact on the price indices, and smaller
than a comparably deflationary price rise of imports in general (R4) or
industrial goods in particular (R6). Export taxes (R5) reduce domestic
prices quite dramatically. These price impacts allow for the important fact
that some prices are set on world markets, so cost changes fall on the primary factors, while other cost changes feed through into final prices. A
standard cost-plus pricing model overstates the impact of these fuel price
changes by about 0.20 of 1 percent or by about one-third.

The table also demonstrates that an increase in all fuel taxes coupled
with an equal yield *decrease* in import duties (i.e., R1-R4) would *lower*
the price level by an appreciable amount (bearing in mind that the tax
changes correspond to only 1 percent of GDP, which can be taken as an
order of magnitude of the expected price change). Thus, if a government
argues against reducing the subsidy on fuel because it fears inflation, the
answer would be to eliminate the subsidy and use the revenue instead for
reducing import tariffs, which will not only lower the price level but reduce
the degree of distortion in the economy.

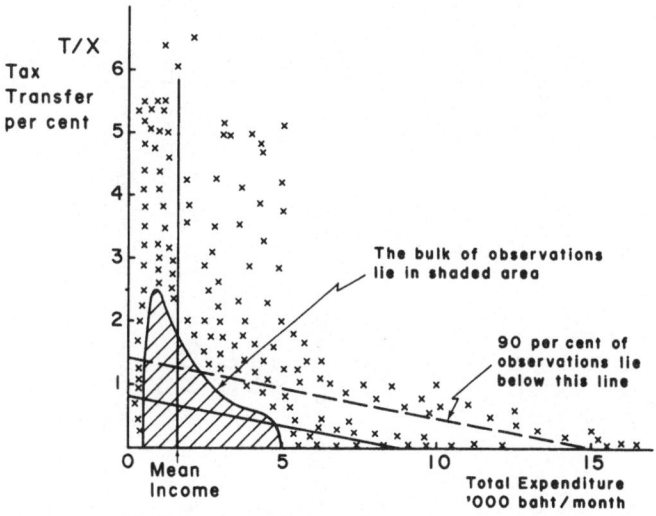

Annex Figure 3.1 Impact of 50 percent kerosene price rise

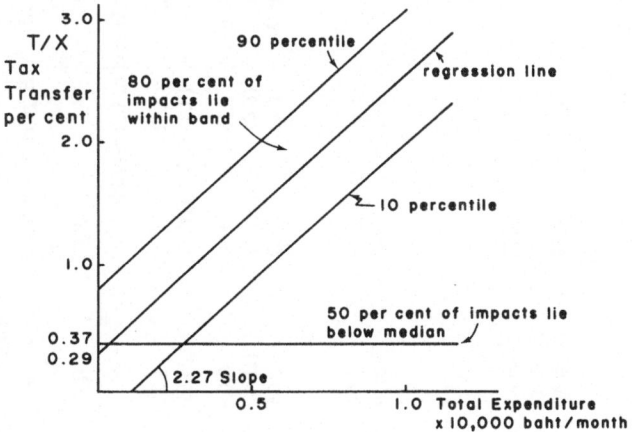

Annex Figure 3.2 Variations in tax transfer for gasoline

Annex Table 3.1 The distribution of net expenditure and income transfers and post-transfer inequality indices for alternative tax schemes, Thailand

| | Tax scheme | | | | | | | |
	All petroleum R1	Gasoline R2	Other R3	Import tariff R4	Export tax R5	Industrial sales tax R6	All petroleum and wage change R7	Industrial sales and wage change R8
A. *Regression equation coefficients for net transfer as a percentage of original expenditure* (all households)								
Real household expenditure (t statistic)	1.12 (22.8)	2.27 (30.9)	-0.27 (4.7)	0.25 (4.5)	-2.43 (8.3)	0.72 (26.7)	1.27 (24.9)	0.90 (25.2)
Household size (t)	-0.02 (6.6)	-0.03 (5.4)	-0.02 (4.8)	-0.006 (1.8)	0.15 (8.0)	0.00 (1.3)	-0.04 (11.4)	-0.02 (8.6)
R^2	0.05	0.08	0.003	0.002	0.01	0.07	0.06	0.06
B. *Distribution of households by net transfer as a percentage of original expenditure*								
Average transfer	0.96	0.66	1.19	0.86	1.36	0.79	1.02	0.86
Median transfer	0.83	0.37	1.10	0.57	-0.24	0.74	0.89	0.78
Deviations between actual and predicted transfers as percentage of expenditure:								
Percentiles: 10th	-0.52	-0.53	-0.60	-0.30	-3.96	-0.43	-0.65	-0.58
Median	-0.12	-0.21	-0.08	-0.03	-1.57	-0.05	-0.12	-0.08
90th	0.69	0.53	0.81	0.45	6.00	0.53	0.80	0.74
C. *Indices of vertical inequality*								
$\epsilon=1.0$	0.232	0.231	0.233	0.233	0.237	0.233	0.232	0.232
$\epsilon=2.0$	0.373	0.372	0.374	0.374	0.379	0.373	0.373	0.373
D. *Indices of overall inequality for* $\eta=2.0$								
$\epsilon=1.0$	0.240	0.242	0.240	0.238	0.278	0.237	0.241	0.239
$\epsilon=2.0$	0.376	0.376	0.378	0.376	0.406	0.375	0.377	0.376

Annex Table 3.2 Inflationary impact of tax reforms, Thailand

	Changes in	
Reform[a]	Producer price index *(percentage)*	Consumer price index *(percentage)*
R1	0.59	0.67
R2	0.37	0.44
R3	0.75	0.84
R4	1.13	0.92
R5	−3.01	−2.03
R6	0.92	0.89
R7	0.42	0.47
R8	0.73	0.69
R1-R4	−0.54	−0.25

Source: Hughes (1983), Table 2.
[a] Reforms defined in the text, p. 83.

NOTES

[1]Strictly, profits must be zero, as they will be with constant returns, or subject to a 100 percent profits tax, for the separation result to hold.
[2]This rather sweeping statement is not uncontroversial, but while it is theoretically possible to argue for some energy taxes or subsidies to supplement the system of direct taxes, no such case has yet been empirically argued. For a further discussion, see Newbery (1981).
[3]This is a loose statement, since cross-price elasticities are also important. One wants to know what happens to total tax revenue if a particular tax is raised. If it induces consumers to switch to less heavily taxed goods, then the tax increase is less attractive than if it induces them to switch to more heavily taxed goods. See Newbery and Stern (1985, Chapter 2).
[4]The National Council of Applied Economic Research (1971) estimated that kerosene used as an adulterant came to 34 percent of officially estimated high-speed diesel consumption in India in 1960, when high-speed diesel fuel had been raised above the price of kerosene.
[5]The qualification "in principle" is required, because to implement the scheme, we need information about the aggregate cross-price elasticities of demand, which in turn requires fitting a consumer demand system for the country. Ahmad and Stern were able to implement the methodology for India and plan to do so for Pakistan and Mexico. Considerable progress in identifying desirable directions of tax reform can be made by just using consumer budget survey data and assuming various values for the cross-price elasticities. In some cases the direction of reform is fairly robust to a plausible range of values of the assumed parameters, in which case one can be reasonably confident of proposing the reform.

REFERENCES

Ahmad, E. and N. H. Stern (1983). *The Evaluation of Different Sources of Government Revenue in India*. Discussion Paper 37. DERC, University of Warwick.

Bhatia, R. (1979). Energy Alternatives for Irrigation Pumping: Some Results for Small Farms in North Bihar. New Delhi: Institute of Economic Growth.

Cecelski, E. and S. Glatt (1982). *The Role of Rural Electrification in Development*. Discussion Paper D73-E. Washington D.C.: Resources for the Future.

Deaton, A. and J. Muellbauer (1980). *Economics and Consumer Behavior*. Cambridge, UK: Cambridge University Press.

Desai, A. V. (1981). *Interfuel Substitution in the Indian Economy*. Discussion Paper D-73B.

Washington, D.C.: Resources for the Future.
Dixit, A. K. and D. M. G. Newbery (1985). "Setting the price of oil in a distorted economy." *Economic Journal*, supplement.
Garnaut, R. and A. Clunies Ross (1983). *Taxation of Mineral Rents*. Oxford, UK: Clarendon Press.
Hughes, G. A. (1983). *The Impact of Fuel Prices in Thailand*. Washington, D.C.: The World Bank.
King, M. A. (1983). "An index of inequality with application to horizontal equity and social mobility." *Econometrica*, 51, pp. 99–116.
National Council of Applied Economic Research (1971). *Consumption Patterns of Selected Petroleum Products*. New Delhi.
National Economic and Social Development Board (NESDB) (1980). *Basic Input-Output Table of Thailand, 1975*. Bangkok.
Newbery, D. M. G. (1981). The Taxation of Oil Consumption. Report commissioned by the Policy Review Unit, British Petroleum. London: 20 July 1981. Manuscript.
Newbery, D. M. G. and N. H. Stern (1985). *The Theory of Taxation for Developing Countries*. Washington, D.C.: The World Bank. (Draft of forthcoming book.)
Sen, A. K. (1973). *On Economic Inequality*. Oxford, UK: Oxford University Press.
Sengupta, D. P. (1979). Energy Planning for Karnataka State: Phase 1, Towards a More Rational Distribution of Electrical Energy. Karnataka State Council for Science and Technology, Indian Institution of Science.
World Bank (1975). *Rural Electrification: A World Bank Paper*. Washington, D.C.: The World Bank.

Chapter 4

OPERATIONALIZING EFFICIENCY CRITERIA IN ENERGY PRICING POLICY

Gunter Schramm

INTRODUCTION

This chapter addresses the question of how the basic economic efficiency rules of energy pricing can be translated into actual market prices, given that there exist important noneconomic objectives as well as practical implementation difficulties. First, it reviews the basic objectives of pricing and discusses the various trade-offs that may be necessary among them. Second, it discusses the problems related to short-run versus long-run marginal cost pricing. Third, it addresses the question of determining long-run marginal supply costs under conditions of economies of scale when potential market sizes differ. This is an issue of considerable practical importance in many developing countries. Fourth, it looks at the related issues of discriminatory and promotional pricing. Fifth, it analyses in some detail the practical questions involved in estimating depletion costs and the importance of the latter for determining minimum economic costs. Sixth, it looks at some of the problems of determining appropriate pricing for petroleum fuels. Seventh, it addresses some special questions related to the opportunity costs of funds which, in turn, determine part of the long-run cost of supply. Eighth, it looks at the reconciliation of economic and financial objectives and, finally, at the problem of dealing with inflation and relative price changes. Two appendixes round out the discussion, one analysing the opportunity costs of oil used domestically under restrictive export quotas, and the other illustrating some of the potential consequences of inappropriately low power prices.

Given the many issues related to the overall topic, the discussion is necessarily selective. In particular, because of the availability of substantive literature on the subject,[1] no attempt has been made to address in any detail the issues of determining the marginal costs of electricity supply systems and setting power tariffs. For illustrative purposes, emphasis has been placed on the pricing of natural gas, an important new fuel for many developing countries.

PRICING, A TOOL OF DEMAND MANAGEMENT

Energy pricing has to be seen in the context of wider energy demand management policies. The overall objectives of the latter are to change demand from patterns that would evolve without management to one that is considered superior on the basis of given policy criteria. Apart from pricing, the major policy instruments are: (1) laws, regulations, and rationing; (2) education and information; (3) policies and regulations affecting the utilization patterns of energy-using equipment and appliances; and (4) direct or indirect subsidies to energy-producing or energy-using activities. The various policy instruments under these four headings can be applied directly to a given energy resource, or indirectly by affecting the cost levels, availabilities, and utilization of energy-using systems, appliances, and machinery. Since these policy instruments are interrelated, their use should be closely co-ordinated for maximum effect.

Among all of them, pricing is a particularly powerful and versatile tool for affecting demand in the long run. In the short run, even sharp changes in prices may have only limited effects on demand but major effects on energy revenue flows instead.[2] Pricing can be applied directly to a given energy source by changing the final price to users. However, it can also be applied indirectly by affecting the prices, costs, or availabilities of energy-using appliances, either through taxes, price controls, direct subsidies, or indirect subsidies provided to energy-producing activities such as tree planting, coal mining, or transportation, or through import controls. Another important means of influencing prices consists of intersectoral cross-subsidies through, for example, lifeline rates for electricity that are compensated for by higher prices to large users, or low-cost pricing for diesel fuel as against high-cost pricing for gasoline. Another form of cross-subsidy consists of country- or region-wide uniform pricing schedules regardless of the specific regional energy delivery costs.

BASIC PRICING OBJECTIVES

The basic objectives that must be considered in energy pricing are (1) economic efficiency, (2) social equity, and (3) financial viability. The efficiency principle seeks to ensure the regulation of prices in such a manner that the allocation of the society's resources to the energy sector fully reflects their values in alternative uses. The equity principle relates to welfare and income distribution considerations. It may result in the charging of differential prices to different users on grounds of basic needs, or in the establishment of uniform prices to specific user groups in spite of differential costs of supply, often justified in the name of regional equity or similar goals. The financial principle suggests that energy supply systems should be able to raise sufficient revenues to remain financially viable, so that continuity and quality of service is ensured. A second objective may be to use the taxation of energy resources as a means to raise required government revenues, either to finance energy-related facilities such as highways, or to raise revenues in general. The latter may be an important consideration in countries in which theoretically more equitable and sophisticated forms of taxation, such as income taxes, do not serve the desired purpose because of ineffective enforcement.[3]

In addition to the three main objectives of energy pricing listed above, there are a number of subsidiary ones which can be important under certain circumstances. One is the objective of energy conservation. The prevention of unnecessary waste is an important goal in general, but there are often additional reasons to conserve certain fuels. These include the desire for greater independence from foreign sources (e.g., oil imports), the goal of reducing environmental degradation, and the need to reduce the consumption of woodfuels due to deforestation and erosion problems.[4] Another objective may be the need for price stability to prevent sudden shocks to energy users and consumers from large price fluctuations. A further objective is the need for simplicity in energy pricing structures to avoid confusing users and to simplify metering and reduce billing expenses. There also may be specific objectives such as the promotion of regional development (e.g., local mining activities or rural electrification, or the support for specific sectors, e.g., export-oriented industries), as well as considerations of other socio-political, legal, and environmental objectives or constraints.

Because these various objectives are often not mutually consistent, a realistic, integrated energy pricing structure must be adaptable to permit trade-offs among them. To achieve this, the formulation of energy pricing policies must be carried out in two stages. In the first stage, a set of ideal prices, which strictly meet the economic efficiency objective, are determined, based on a consistent and rigorous economic framework. The second stage of pricing then would consist of adjusting these efficiency prices (established in the first step) to meet all other objectives and constraints.

Given the many noneconomic objectives that will ultimately affect the level of prices, the question might be asked whether it makes sense to establish a set of efficiency prices in the first place without simultaneously taking into account all other objectives. The answer is: yes, it does. First of all, it is useful to know by itself which set of prices will reflect the least-cost solution of providing energy. Second, only if it is known what this set of prices is can an assessment be made of the costs that the various other distributional or nonquantifiable objectives may have. The set of efficiency prices provides something of a yardstick that can be used to measure the consequences and economic costs of introducing these other goals, even if it cannot tell what the real value of these other goals is. In some cases, the costs imposed by them in terms of losses in efficiency are unacceptably high. In others it might even turn out that the proposed beneficiaries of a specific, distributional objective may voluntarily opt for some other form of compensation instead if they find that such compensation makes them better off (e.g., low-cost electricity for cooking instead of subsidized kerosene). If, on the other hand, such distributional or non-quantifiable goals are directly incorporated into our set of pricing, there is no way of measuring resulting losses in economic efficiency. Knowing the latter will greatly facilitate evaluations of potential trade-offs between the multidimensional objectives that invariably form part and parcel of any price-setting decision.

Efficiency objectives

A fundamental consideration in energy pricing is that prices to users should reflect the full, long-term, marginal social opportunity costs of their use. In establishing this economically efficient price, the real rather than the financial costs of the resources utilized must be used. This means that shadow prices should be applied whenever real values diverge from market prices. Usually, the most important of these is the shadow price of foreign exchange. With the steep rise in the cost of energy imports and their adverse effects on the balance of payments of most countries, the shadow price or premium above the official rate of foreign exchange may be substantial and may even increase over time as a result of the increasing costs of energy imports. On the other hand, the shadow price of labour may be important only in cases of labour-intensive activities, such as plantations for fuelwood, alcohol production, or programmes to introduce more efficient cooking devices in rural areas. Separate shadow prices may also be needed for establishing the real value of scarce public investment capital. However, while the principles and need for shadow pricing are widely accepted and understood,[5] empirical information about the magnitude of the various shadow pricing coefficients, is usually lacking. This shortcoming must be remedied because the distortion between market and shadow prices may be substantial, with the result that energy users charged on the basis of market prices may receive undeserved subsidies that must be paid for by some other sector of the economy.

The base price of any energy resource is determined by its long-run marginal costs. Marginal costs establish forward-looking prices. Such prices reflect the real value of all additional resources that must be utilized in order to make another unit of energy available. These marginal costs include costs of the investments that are needed to supply the additional units of energy. If prices are below this level, there will be a net economic loss in the long term because energy use will be higher than it would be otherwise and is justified on the basis of real resource costs.

The second important principle is that energy prices should reflect the value of the energy resources consumed in their next best alternative use. In the case of easily transportable petroleum products, for example, the next best alternative use is usually given by the export price (f.o.b.) or import price (c.i.f.) of the petroleum product, adjusted for any quantity or quality differential and additional transportation costs. Export and import prices at a given location may differ by a significant margin. For example, fuel oil delivered to Chittagong from Singapore may cost US$160/ton at today's prices, while the f.o.b. value of surplus fuel oil, exported from the same port, may be only US$140/ton. Net differences in inland locations, say, at a refinery in Assam, may be much higher, perhaps around US$45 to US$60 per ton. The same applies to other energy resources such as natural gas or coal. For these, moreover, transport costs usually are considerably higher per unit of energy, with the result that their economic net value in the next best alternative use is correspondingly lower.

Third, energy prices should reflect all external costs (or benefits, if any). For example, external benefits from certain energy uses such as increased use of kerosene or electricity instead of firewood may reduce the overcutting of timber resources for fuel and thereby reduce erosion, recurrent flooding, or reservoir siltation. Hence the question of external benefits is an important issue for the pricing of fuels such as kerosene or liquefied petroleum gas (LPG) in badly eroded areas, or for the evaluation of the total benefits from reforestation projects.

Typical external costs of energy usage are pollution and congestion. Pollution costs, mainly through air pollution from high-sulphur fuels, may be significant in large metropolitan areas such as Bangkok, Bombay, or Singapore. Their main effects are related to health. Congestion costs, on the other hand, are mainly economic costs. They consist of three major components. The first is the additional amount of fuel consumed by vehicles held up by congestion; the second consists of the time lost by drivers and passengers; and the third of the additional costs of less than optimal utilization of the vehicle themselves (the costs of waiting plus the costs of fewer ton-miles or passenger-miles per vehicle). Bangkok provides a visible example of the magnitude of these costs.

Fourth, prices of exhaustible, domestic energy resources such as crude oil, natural gas, coal, or hydropower storage capacity subject to siltation should reflect their foregone current or potential future net value. The latter is usually called "user," or "depletion" cost. It measures the future net economic value of exhaustible energy resources that are used now and must be replaced by higher-cost alternatives later.

The economic value of depletable resources is determined by five types of opportunity costs. The first consists of the long-run marginal costs of supply, which include exploration, development, processing, transmission, and distribution costs. The second represents the foregone future net value of the resource once it is depleted and must be replaced by alternative resources. These are the "user" or "depletion" costs. The third is determined by the net value of the resource in alternative uses (as indicated by its f.o.b. export price, net of production and delivery costs, and depletion allowance). The fourth represents the net value of the resource as a current substitute for other energy resources, net of all differences in delivery and usage costs between alternative fuels. The fifth is determined by the net value of the resource in uses that would not occur if alternative, higher-cost energy or feedstock materials had to be utilized. Examples are fertilizer production or liquefied natural gas (LNG) exports whose viability depends on prices below those of alternative fuels.

The first two of these opportunity costs — the long-run marginal supply and the user costs — are additive and represent the basic economic costs of the resource. They determine the minimum price that has to be charged for the resource. If lower prices were to be charged, net losses to the economy would be incurred. The other set of opportunity costs must be higher than the sum of the former in order to produce economic net benefits. They also determine the economic ceiling prices that could or should be charged to users. While, in many cases, higher prices could be charged if the sale or importation of substitutes at lower prices is prevented, this would be economically inefficient, even though the users' willingness to pay might be high enough to sustain such price levels.[6]

An important issue in the determination of the net value of a resource is that there will generally be a number of different markets where the resource could be sold. However, the unit value of the resource to these users is likely to differ substantially; it may also differ for a given user.[7] These differences depend on the specific characteristics and sizes of the respective markets. Because of such differences, it would be wrong to conclude that the value of existing, depletable resources in the ground is determined by their highest-value use (whose market share might be quite limited relative to available supplies). While resource allocation rules should generally try to fill the requirements of the highest-value markets first, subsequent allocations should follow the common-sense, economic optimization principle that supplies should be made available in declining order of net benefits until either all available resources are fully committed or, until, at the margin, the value of the last unit committed is just equal to its economic costs (that is, until it equals the costs of production plus user costs). It must also be noted that the latter costs increase with any additional allocation due to the effect of such added allocations on reserve/production ratios.

Equity objectives

Socio-political or equity arguments are often advanced in favour of subsidized prices for energy, especially where the costs of energy are high relative to the incomes of poor households. Economic efficiency arguments based on externality effects may also be used to support subsidies (e.g., cheap kerosene to reduce excessive fuelwood use to prevent deforestation and erosion). Two issues should be noted in this context. First, low-priced, so-called "lifeline" rates may deviate markedly from economic efficiency criteria. Second, the amount of the subsidy that is to be made available through the lifeline rates must be carefully monitored so that either the revenue of the subsequent higher-priced blocks balances the losses incurred or a sufficiently high subsidy is paid by the government; otherwise, the financial viability of the supply organization will be jeopardized.

Sometimes, initial large subsidies and temporary losses may be justified on economic grounds if it can be expected that demand from the new users will eventually increase sufficiently to ensure adequate capacity utilization and recapture of the initial subsidy.[8] Many types of pricing policies can be used to assist or subsidize specific user groups or societal sectors. The most common forms of subsidies consist of differential product prices and excise tax levies on specific products. At present, kerosene, LPG, diesel fuel, and fuel oil are subsidized in many countries, usually on the ground that these products are essential inputs to specific users that must be made available at low prices. However, as will be shown below, such arguments are generally fallacious. Though attempts are usually made to sell other petroleum products, mainly gasolines, at high prices, the accounts often do not balance and the additional revenue raised is insufficient to cover the losses from the subsidized products. A further problem resulting from heavy cross-subsidies is that the demand for subsidized products (e.g., kerosene and diesel) often outstrips the demand for other refinery output (e.g., gasoline and fuel oil). As a result, several countries are forced to re-export the latter at substantial costs, while importing refined products at premium prices.

Financial objectives

Two major financial objectives must be considered in setting energy prices. The first is the financial viability of the energy supply organization, while the second relates to the general revenue goals of the government. The financial principles are often embodied in criteria such as target financial rates of return on revalued assets, or acceptable rates of contribution towards the costs of future investment programmes. Providing sufficient revenue flows to energy supply organizations, whether they are publicly or privately owned, is of major importance for maintaining efficient and reliable operations (although the meeting of financial targets is only a necessary but not a sufficient condition to meet this goal). Without sufficient revenues, day-to-day operations will suffer, maintenance will be neglected, plant and equipment will deteriorate, and capable staff will leave. The results are unreliable energy supplies which are far more costly to an economy than high energy prices. This has been illustrated by the analysis of the cost of unreliable power supplies contained in Annex 4.2.

Taxation of energy supplies has been found in many countries to be a cost-efficient device to collect needed governmental revenues if the demand for such energy resources is relatively inelastic (because higher prices do not lead to significant changes in consumption, such taxes do not have a major distortional effect in terms of economic efficiency losses).[9] A subobjective for raising revenue through energy taxes might be to cover all or part of the costs of energy-related government expenditures on, for example, roads.

SHORT-RUN VERSUS LONG-RUN MARGINAL COST PRICING

It is argued here that the appropriate base for determining efficiency prices are long-run marginal costs. While this view is widely held, it is challenged by at least some economists, who argue that short-run marginal costs should be used instead.[10] As will be shown here, for many energy price determinations the difference between short- and long-run marginal costs is more apparent than real and becomes important only in cases in which lumpy, nonrecurrent capital expenditures have to be accounted for.

One important aspect for short-run marginal costs pricing that is generally overlooked is that in cases of capacity shortages short-run marginal costs become really discontinuous. What this means is that even substantial increases in prices will not bring forth new supplies. All that these higher prices can do is choke off part of the existing demand until equilibrium is reached between willingness to pay (i.e., demand) and available supply. However, such an enforced equilibrium also eliminates any market signal that new investments are needed, unless short-run marginal costs have risen to such a level that they actually can cover the instantaneous costs of such investments.

To understand the essential difference between short-run and long-run marginal costs, it is useful to review briefly their definition and meaning.[11] Marginal costs are defined as the net change in total supply costs resulting from an incremental change in output. This means that in the short-run only variable costs (i.e., the costs of those inputs that vary with changes in output) form part of the marginal cost accounting framework. Because the fixed costs of existing plant (e.g., capital equipment, buildings) remain constant, they are netted out and ignored in the determination of marginal costs. Such pricing is correct from the viewpoint of economic efficiency because prices that reflect marginal costs are equal to the net opportunity costs of resources at the margin needed to bring forth the additional supply.

However, the strict application of such prices is appropriate — or feasible — only in a static world in which there is no change, in which demand remains constant or declines, in which no lumpy investment is ever needed to increase capacity or to replace worn-out equipment or depleted resource deposits.

Practical difficulties with this pricing approach are encountered when new investments are needed. These usually are lumpy and require large amounts of resources that must be committed first before any additional output can be produced. Since the costs of such investments prior to their irrecoverable commitment are variable, they have to be included in the calculation of overall marginal costs. However, as soon as they have been made they become "sunk" costs so that they no longer affect decisions at the margin. As a consequence, marginal costs again fall to the incremental level of operating (i.e., variable) costs, and investment costs once again are ignored.

The amplitude of these price fluctuations resulting from such "before" and "after" considerations in typical developing country energy supply systems would be huge, if the costs of the additional, required capital investments were to be charged to consumers at the time new investments have to be made. Price fluctuations of such magnitude would be unacceptable for any economy. They would certainly be highly disruptive to any energy-cost-sensitive activity such as cement, pulp and paper, or steel production, or transportation. They would also be unacceptable to domestic consumers. Economic as well as political considerations would rule out the adoption of such pricing patterns.

This means that modifications of the simple, short-run marginal cost pricing principle are needed. These modifications should meet three criteria: First, they should maintain the basic integrity and advantages of marginal cost pricing, aiming at the equivalence of willingness to pay to incremental cost of supply at the margin. Second, they should assure that all supply related costs are borne by the respective consumers. Third, they should maintain reasonable long-term price stability or price predictability to facilitate forward planning of energy-use related investments.

Two possible alternative approaches offer themselves. The first is to utilize some form of two-part tariff, which would consist of a fixed periodic charge (or one-time connecting charge) reflecting capital costs and another reflecting the short-run marginal cost of the energy supplied. Such tariffs have been particularly recommended for situations in which peak-load capacities are needed. However, two-part tariffs can be utilized only for those energy resources that depend on fixed connections with metering devices (e.g., electricity and natural gas). They would be impractical for all energy resources which could easily be resold outside formal market channels. The latter applies to all petroleum products, natural gas liquids, coal, charcoal, and wood.

Two-part tariffs would be increasingly more inefficient the higher average capital costs are relative to operating costs. This is so because higher average capital costs would lead to high fixed charges and low energy costs. A potential energy user would have to either pay the fixed charges or do entirely without this source of energy. Once he agrees to pay, the fixed charges would no longer affect his consumption pattern. Only the energy costs would be relevant to his decisions. If the latter were low, wasteful usage would likely result. This waste, in turn, would result in higher growth rates, which would require larger and more frequent additions to capacity. But capacity costs, once again, would not affect energy use, creating a vicious cycle of rapidly rising, economically unjustified energy-use patterns. Also, such tariffs would tend to exclude the poor since they could not afford to pay the high fixed charges. Hence, two-part tariffs for the purpose of financing all capital costs do not appear to be useful except in cases in which short-run marginal costs are a substantial proportion of long-run marginal costs.

The other alternative for dealing with indivisibilities would be to utilize a forward-looking averaging approach. The costs of forthcoming investments (i.e., the marginal investment costs) would be spread over an appropriate period, usually the life expectancy of the asset or, sometimes, its financing period. These levelled-out capital costs, annuitized at the appropriate interest rate, would be divided by the energy units supplied per year and added to the marginal operating costs. The total unit charge would then reflect long-run marginal costs, in contrast to the short-run marginal costs defined above.

Including this annuitized capital cost charge in the marginal cost price structure actually is a vitally important signal to an energy consumer of the real costs of his consumption. With growing demand, each additional unit consumed encroaches upon existing capacity and raises the spectre of additional future investment costs. The levelized capital costs and charges, therefore, are nothing but a measure of these future costs. What we can conclude, then, is that long-run marginal costs represent the true measure of the actual economic costs of supplying additional units of energy.

Newbery argues that the difficulties of short-run marginal cost pricing can be dealt with by offering contracts of varying length during which an agreed quantity of electricity is sold at an agreed stable price. Variations in consumptions above or below this contracted amount would be priced at the spot price, or the short-run marginal cost.[12] He does not say, however, how this "stable price" for long-term contracts is to be calculated. However, having defined short-run marginal costs as the spot price, the contract price presumably is to be based on long-run costs, i.e., presumably long-run marginal (rather than average) costs. Selling the remaining temporary surpluses in a spot market at short-run marginal costs, of course, makes eminently good sense and does not violate the principle of long-run marginal cost pricing.[13] Such "spot markets" for electricity supplies are well known and widely used. They consist of the sale of so-called "secondary" or interruptible energy to either neighbouring systems, or to users who can either switch to alternative sources of supply when they are cut off (e.g., to auto-generation facilities) or who can do for some time without power (i.e., cold-storage facilities). Off-peak rates for such types of uses are widely known and used.

EFFICIENCY PRICING WITH DECLINING LONG-RUN MARGINAL COSTS

The long-run marginal supply costs of specific energy systems may change significantly, depending on differences in initial demand, rate of demand growth, and capacity utilization of usually lumpy investments, such as extraction, transmission, and distribution facilities. For example, market studies in a developing country for a newly discovered gas deposit showed that gas should definitely be used by industry, implying a certain system configuration and size for the gas supply system. However, there was also the possibility of using gas for power generation, in competition with low-cost hydro, and, possibly, in transport, displacing petroleum fuels. If these additional markets were to be served, different production and transmission systems would be needed and load built-up and capacity utilization rates would differ. But such supply facilities are subject to considerable economies of scale. For example, in this specific case, average costs per unit of gas supplied were found to fall by some 60 percent, and marginal costs by almost 85 percent, as pipe diameters doubled and resulting throughputs increased by a factor of five. Increased markets, therefore, in this case, meant substantially reduced long-run marginal costs of supply. Such economies of scale are typical for less than technically optimum-sized energy supply systems. To reduce unit costs, therefore, a major concern of energy supply systems must be the rapid development of optimum-sized markets to reduce the unit costs of supply. This can mean, however, that supplies from a given energy resource should not only be sold in premium markets but also in those in which its value in terms of available substitutes would be considerably lower. The basic test to be applied in such cases would be to check that the long-run marginal costs of supply plus depletion costs (in the case of a depletable resource) is equal to or lower than the price that this marginal user would be willing to pay. In the country to which the above data refer, for example, electric power could also be obtained from low-cost hydro sites. However, once the very low marginal costs of increased capacities of the gas transmission facilities were taken into account, the costs of electricity from gas-fired plants turned out to be lower than those from hydro. In some countries, such as Thailand, potential scale economies can have a significant effect on the question of whether or not future power developments should be based on lignite, on imported coal, on domestic gas, or added hydro developments.[14] However, while long-run marginal extraction and supply costs may decline with increased output, per unit depletion costs inevitably rise as rates of output increase because with increased output the time to resource exhaustion is shortened. This increases unit costs. On the other hand, the marginal value of a resource in given uses often decreases as supply is increased.[15] What this means is that the economic unit value of a depletable resource is subject to changes in several economic parameters that move in different directions as output increases. Detailed and often complex calculations are needed to estimate the resulting net economic values, as well as optimal rates of extraction over time.

THE ECONOMICS OF DISCRIMINATORY PRICING

Price discrimination can be defined as the charging of prices to selected groups of customers that differ by different margins from the social long-run marginal costs of supplying them. Such differential pricing is common. It should be noted, however, that not all price differentials automatically qualify as discriminatory pricing. Differences in the quantity, quality, timing, and location of deliveries will result in differences of marginal costs; these should be appropriately accounted for in the setting of prices and do not represent price discrimination. True price discrimination is widely practiced whenever it is possible to differentiate between user groups. It is common in pricing schedules for electricity and natural gas systems. These can easily discriminate among customers because they supply through individually metered connections. Discrimination is less common for other types of energy supplies because of the difficulties of preventing resales (e.g., petroleum products).

There are many reasons why price discrimination is practiced so widely. Factors are income distributional objectives, attempts to foster economic developments through low energy prices to specific sectors, or simply outright political pressure. However, there are a number of situations in which price discrimination can be an important tool in bringing about the economically most efficient development of given energy resources for specific markets. Such conditions can arise when new, potential additions to supply exhibit substantial indivisibilities and are very large relative to existing markets. Hence, while their average unit costs may be attractively low at full production, market-size limitations may be such that the unit costs would be unacceptably high at lower rates of output. Situations like this are common, particularly in developing countries in which market size often is a major constraint. As a result, many of them are saddled with low-volume, small-scale energy supply systems whose unit supply costs are far higher than those of potentially available, optimum-scale systems. A typical example is provided by Nepal's ample hydropower resources whose low-cost sites are far too large to be suitable for the domestic market. As a consequence, small-scale, high unit-cost projects have to be utilized.

There are two options — both involving discriminatory pricing — that can sometimes be used to overcome these limitations imposed by market size. If the relevant domestic market is relatively inelastic and the quantitative difference between market demand and a new project's potential supply is not too large, discriminatory pricing schedules may raise enough total revenue to cover the long-run marginal costs. This case has been illustrated in Figure 4.1. As can be seen, the total market demand schedule lies below the long-run marginal cost schedule at any given single price. However, discriminatory pricing, with P_1 charged to user group 1 which purchases OQ_1 at that price, P_2 charged to group 2 purchasing Q_1Q_2, and P_3 charged to group 3 purchasing Q_2Q_3, results in average revenues equal to P_a which are equal to the long-run marginal cost at F. It should be noted that under such pricing schedules, it may well be necessary to sell substantial quantities of output at prices below the long-run marginal cost, although never at prices below short-run marginal costs. If the long-run marginal costs of the project under consideration are lower than those from any competing alternative, aggregate economic welfare will be maximized by choosing this project, despite the need for using discriminatory pricing in order to make it economically and financially feasible.

The other potential alternative is to search for additional markets for the project's surplus output, even if it means that this output must be sold at substantially lower prices than those charged in the primary domestic market. Such additional markets can sometimes be found through exports. In Bangladesh, for example, supply of natural gas from the eastern fields to the western part of the country across the forbidding Jamuna River would be economically feasible only if a large export market for gas could be found in neighbouring India. Without such an additional market justifying the huge expense of a gas pipeline river crossing, delivered gas costs in western Bangladesh would be far higher than those of imported petroleum fuels.

New markets or demands sometimes can also be created by attracting energy-intensive activities to the vicinity of a new project. Examples are Ghana's Volta River power development, which depended on a new, exclusively export-oriented aluminium smelter to market most of its initial power output, or the Aswan Dam in Egypt, which sells substantial portions of its electric production to an aluminium smelter and an energy-intensive fertilizer plant. However, the history of both of these undertakings also illustrates some of the potential, long-term problems of such ventures. Both Ghana and Egypt are now short of electric power and have to develop new, high-cost sources of electric supplies because substantial quantities of the initially available, low-cost power are committed and must be sold at very low prices under long-term contracts.

PROMOTIONAL PRICING

Project indivisibilities or economies of scale and the lack of temporary, alternative markets for surplus outputs may make it desirable to use promotional pricing schemes, resulting in temporary financial losses, in order to attract a larger number of customers quickly, stimulate greater consumption by existing ones, and thereby expand project utilization at a faster rate. Another reason for promotional pricing of a specific energy resource may stem from the desire to reduce the consumption of alternative energy sources whose use entails substantial negative externalities, as, for example, the overcutting of forest resources.

Promotional pricing is defined as the temporary underpricing of energy supplies to selected customer groups at levels below the long-run marginal cost. A typical example is provided by rural electrification networks. For these, distribution costs are usually high because of long distances and low load densities. Technical considerations, however, require minimum investments in terms of line voltage, number and structural strength of distribution poles, conductor size, etc. Almost all of these initial costs, including those of meter reading, are fixed costs, at least up to network capacity. Hence unit supply costs are inversely related to sales, and increased sales would reduce the long-run marginal cost accordingly.

Energy use depends on the use of energy-consuming appliances; in the case of electricity, light bulbs, refrigerators, flat irons, hot plates, etc. If electricity is priced at low rates, more users may be willing to invest in such appliances, and more users may be willing to sign up to become paying customers. This would increase unit sales per customer, as well as the number of customers per given line, thus reducing total unit costs. For example, in Thailand, which maintains a vigourous rural electrification programme, usually about 40 to 50 percent of all households will sign up initially for electricity supplies when the distribution lines first reach a village. This rate increases to between 75 to 80 percent after three to five years. Average per household consumption in newly electrified villages is about 40 percent of the average consumption in all rural areas combined.

Hence it may be reasonable and economically more efficient to price energy supplies not at the present low-volume, high-unit, long-run marginal cost, but at the expected average lifetime long-run marginal cost that is based on a more rapid, immediate load build-up. For example, in a system that reaches its design capacity after a load build-up of 10 years and has a life expectancy of 25 years, average long-run marginal unit costs will be about 10 percent higher than those in a similar system that reaches its design capacity after a load build-up of only five years.[16]

An important consideration in adopting such a pricing scheme must be that the financial resources of the supply organization have to be sufficiently large to cover the initial financial losses incurred and that demand forecasts are realistic.

ESTIMATING DEPLETION COSTS

Positive depletion costs will be incurred in the utilization of an exhaustible energy resource if the costs of the next best fuel that has to be utilized after exhaustion are higher than the long-run marginal supply cost of the depletable resource. The magnitude of the depletion costs per unit of extraction varies with (1) the delivered net-cost differential in use between the depletable resource and its substitute, (2) the production/reserve ratio of the resource (i.e., the time to exhaustion), and (3) the economic rate of discount (inversely).

By definition, the marginal supply plus depletion costs must be lower than the comparable costs of the replacement resource. However, the combined costs of the two increase over time at the rate of discount until they reach equality with the costs of the replacement source at the time of exhaustion (or complete resource commitment). This has been illustrated in Figure 4.2 which shows the relationship between marginal extraction costs (assumed to be constant for simplicity of exposition) and replacement costs at constant prices, as well as with rising relative prices. As can be seen, marginal costs plus depletion costs equal to $MC + U'$, or $MC + U''$, respectively, rise between time t_0 and time t_m (the time of resource exhaustion) at the rate U' or U'', reaching the replacement costs P' or P'' at the time of resource exhaustion.

If the net differential in marginal supply costs between the resource and its replacement is low, and/or the time to exhaustion is long, and/or the rate of discount is high, the depletion cost allowance will be low.[17] Table 4.1 indicates some representative cost ranges, based on assumed, net marginal costs differential of US$1 per unit between the depletable resource and its next best substitute. As can be seen, at a rate of interest of 4 percent and a life expectancy of 10 years, the current, present value of depletion per unit of extraction is equal to U$0.50; with a reserve life of 20 years it falls to US$0.23; at a rate of interest of 12 percent, however, unit depletion costs are only US$0.15 with a resource life of 10 years and a negligible US$0.02 with a resource life of 30 years. What this means for practical resource allocation, as well as pricing decisions, is that in countries with large, depletable resources and/or high opportunity costs of capital, the *in situ* value of these depletable resources is low; for such countries a strategy of rapid development and use may well be optimal, even if prices obtainable are relatively low, or utilization costs are relatively high.[18]

Complications in calculating depletion costs arise when it becomes necessary to assess them for a large-scale resource allocation that is independent from the depletion costs created by other uses. Such situations arise, for example, in the case of large, long-term export contracts.

In situations in which the quantity assigned under a proposed contract is large relative to the size of the deposit, the time path of exploitation to exhaustion will change. Also, the marginal supply costs to other users may be affected and might increase or decrease. It would decrease if common supply facilities can be used and economies of scale are present. However, it could also happen that the new use crowds out some other potential users. The limited supplies remaining for other uses in such situations may force a downsizing of transmission and distribution facilities because of lower consumption and lower throughputs over time. The net effects of these changes on systems supply costs have to be accounted for to get a true measure of the actual depletion costs caused by the new, large user.

To deal with the totality of these changes and compute their magnitude, it is necessary to calculate the net economic costs of not being able to utilize the contracted quantity of the resource in alternative uses. This value represents the net depletion costs of the project.[19] Figure 4.3 illustrates this point. Without the project, "normal" consumption would grow from AB today to TC at a time T, when the maximum allowable rate of production would be reached. Thereafter, production would remain constant until exhaustion at a time Tx.[20] The proposed project would require the allocation of a quantity of gas equal to area EFC'G'H'C'', to be utilized at constant annual rates equal to EF (=C''C' or H'G').

The new project is assumed to start operation in year t'. Therefore, in that year, total gas production has to increase from t'E to t'F. Thereafter, annual production rates increase in accordance with the projected growth of demand by all other users until the maximum allowable annual rate of production is reached. This occurs at time T', instead of at T in the "without" case. The year of final exhaustion of the reservoir is advanced from year Tx to year T'x.

The net reduction in total domestic consumption over time, which, of course, is equal to the amount of gas assigned to the project, is shown in the diagram by area C''C GDD'H'. To calculate the additional user costs caused by the project, it is necessary to find the differential in total energy supply costs to other uses with and without the project, properly adjusted for differences in gas supply systems costs.

Under certain conditions, this cost element may be quite large relative to total costs. This is illustrated by data from a proposed gas development project in a developing country. An international manufacturer wanted to use gas from a newly discovered deposit as a feedstock for an export-oriented chemical plant. At the time the negotiations took place, no domestic gas uses had developed. However, market studies showed that there existed a modest potential for domestic gas utilization by industry and for power generation, with the gas substituting mainly for high-cost, imported petroleum products. Table 4.2 summarizes the major data relevant to the case.

The gas deposit contained some 700 billion cubic feet (ft^3) and was located close to tidewater and harbor facilities. The foreign manufacturer asked for a contract for some 500 billion ft^3 for use as feedstock. The offered contract price was US$0.75 per 1,000 ft^3. On the other hand, market studies indicated that the net value of the gas in domestic uses was much higher, ranging from US$1.60 to US$3.20/1,000 ft^3 exclusive of long-run marginal supply and depletion costs, with the range depending on the growth of demand and specific types of uses.

However, the potential domestic market was small relative to the size of the gas deposit. Depending on various projections of demand, the gas deposit, used for the domestic market only, could have lasted from 32 to 50 years. Because of these long time horizons, estimated depletion costs were quite low, amounting to only about US$0.17 per 1,000 ft^3 (with 32-year life) or 14 percent of total domestic gas supply costs.

With the chemical plant included, however, maximum annual domestic gas consumption would have had to be limited to 12 billion ft^3 instead of 35 billion ft^3 in the "without" case. At projected, most likely rates of domestic demand growth, the available gas in the "with plant" case would have been exhausted after 20 years in spite of the lower consumption rates. With this scenario, average depletion costs for all users increased to US$1.02/ 1,000 ft^3, or to a level well in excess of the offered gas contract price for the chemical plant.

To determine the net economic depletion costs imposed by the proposed project, the foreclosed domestic consumption profile for the 500 billion ft^3 of gas to be contracted had to be determined. This showed that the specific depletion costs attributable to the proposed chemical plant would have amounted to some US$1.19/1,000 ft^3, or some 82 percent of the plant's total estimated supply plus depletion costs of US$1.45/1,000 ft^3.[21] Hence, to cover the chemical plant's economic opportunity cost to the country, its average delivered gas price had to be at least US$1.45/1,000 ft^3,[22] almost double the offered contract price.

THE ECONOMIC VALUE OF DOMESTIC PETROLEUM RESOURCES

For countries with domestic crude resources, ascertaining their economic value appears quite simple and straightforward. It is the f.o.b. export price, adjusted for special, export-oriented storage and handling costs. However, while this is true for countries that can sell all of their oil production freely at prevailing world market prices, it does not apply to members of the Organization of the Petroleum Exporting Countries (OPEC) that are subject to export (but not to production) quotas. For such countries, and in such situations, the value of a barrel of oil consumed domestically is given by the future, rather than the present value of the barrel sold abroad. If, for example, export restrictions are projected to continue for a period of ten years, and domestic production capacity is projected to be in excess of quotas for this period or longer, the net value of a barrel with, for example, a current f.o.b. export price of US$28 would be no more than US$10.80 if it is consumed domestically. However, if the quota restriction applies to production, rather than exports only, today's net opportunity costs would be determined by the current export price.[23]

THE ECONOMIC VALUE OF IMPORTED PETROLEUM FUELS

For countries that depend entirely on the importation of refined petroleum products, the calculation of their economic value is quite straightforward. It is the c.i.f. costs of these fuels, broken down by category, plus all handling, storage, transportation and distribution costs, properly shadow priced, if necessary.

For countries that are petroleum importers but refine some or all of the imports in domestic refineries, the issue is more complicated. Refineries typically use a single input,[24] whose economic value is equal to its shadow-priced c.i.f. costs. However, a refinery's output consists of a wide range of products that are jointly produced, ranging from liquid petroleum gases to light and middle distillates to fuel oil, sometimes asphalt, and perhaps various lubricating oils and greases. Because most of the production of this wide range of outputs is joint, product-specific costs cannot be allocated, except on an arbitrary basis. To determine ex-refinery prices, therefore, arbitrary rules have to be applied. The only fundamental principle must be that total revenue (including f.o.b. revenue of re-exports, if any) must be equal to the shadow-priced import costs of the crude, plus the economic costs of the refining operation, plus storage and handling.

Because of this joint production and cost allocation problem, it is necessary to find the value of each individual product in some other way. One of them is by evaluating it in terms of its economic opportunity costs, rather than its cost of production. These economic opportunity costs can be found by calculating the separable f.o.b. price of each individual product if it is a surplus product at the margin and has to be exported, or the c.i.f. import price, if refinery outputs are insufficient to supply demand. Care must be taken, however, to account for artificial market restrictions that may depress consumption for given products.[25]

Special problems arise if crude imports plus refinery costs are higher than comparable prices of imported products. This could be the case if a refinery is small, outdated, or inefficient and if the difference between the shipping costs of crude and products is not too high.[26] In such cases, decisions have to be made either to close down the refinery or to modernize it to become competitive.

THE ECONOMIC OPPORTUNITY COSTS OF FUNDS: A DIGRESSION

So far the discussion has implicitly assumed that all economic values would be calculated by applying the appropriate real rate of discount. This rate, given the chronic capital shortages of most developing countries, is generally estimated to be quite high. The World Bank, for example, usually requires minimum rates of 10 percent in real terms for project evaluations;[27] the Planning Commission of Bangladesh insists on a rate of 15 percent. Applying these rates is appropriate if the funds to be utilized have true opportunity costs (i.e., if they could be used elsewhere in the domestic economy if the specific project does not materialize). This applies also to funds from multilateral or bilateral donors, regardless of the actual interest rate charged, as long as these funds are "fungible."

There are a few special situations, however, in which the "fungibility" arguments do not apply. One is given in the case of private investment capital that would flow into a given country only for the project under consideration. Private funds for oil and gas exploration and development are prime examples. In such cases, the initial, basic opportunity costs of these foreign funds is equal to zero. However, these investors expect to be compensated through the flow of future dividend and interest payments, as well as depreciation charges. These future outflows of foreign exchange funds represent the real economic opportunity costs to the country. They have to be appropriately discounted to the present and accounted for as the opportunity costs of these investments.

The second case of potentially low opportunity costs arises in cases of foreign aid that is tied exclusively to a specific project and that would not be made available under any circumstances for some other, alternative investment. Such examples are not unknown. A 210-megawatt outside-financed, gas-fired steam power plant in one of the countries of the region is an apparent example of such a specific, project-tied, zero-opportunity-cost project.[28] To account for the true economic costs of such projects, their nominal interest rate and agreed-upon repayment schedule should be used as the appropriate measure of their economic opportunity costs.

RECONCILING ECONOMIC AND FINANCIAL COSTS

Economic criteria should provide the foundation for setting prices to consumers of energy resources. However, economic costs may deviate significantly from financial or market costs. Setting prices on the basis of economic criteria only, therefore, could bring about large windfall gains to some, or substantial financial losses to others. In situations in which market prices, foreign exchange rates, and bank lending rates are controlled by the government and set at values different from those that would prevail under unrestricted market conditions, the levels of prices required to cover long-run marginal economic costs are likely to be higher than those required to cover actual market costs of supply. For example, with an overvalued exchange rate, the economic shadow-priced imports would value imported inputs higher than the actual nominal costs. Regulated interest rates may be lower than those indicated by economic opportunity costs calculations, whereas local labour costs may command a lower economic value than those indicated by the wage rate.

If an energy supply organization confronting these different sets of economic and market prices is government owned, it could be argued that the likely windfall gains resulting from setting prices at economic costs, rather than market-based costs, make little difference, because the government, as the only shareholder, will ultimately benefit from the excess profits accumulated. However, what it also means is that the energy supply organization would accumulate large amounts of surplus funds that it might decide to use internally rather than promptly turning over to the government.[29] For this reason alone, it would be far more appropriate for the government to impose excise or other taxes on the energy supply organization that would siphon off this differential between economic and market-based costs and transfer the funds to the treasury which, after all, has to bear the brunt of the real costs of manipulated market prices.

The opposite problem can also arise, however. Long-run marginal cost-based prices can easily result in financial losses if future economic costs are projected to decline significantly from current levels. This is not uncommon in developing countries with energy supply systems subject to potential economies of scale.

For example, the long-run average incremental costs of the power system in one African country were estimated recently at US$0.08/kWh. This calculation took into account the system expansion plans to the year 2000 and assumed that a projected long-run, high growth demand scenario would prevail. This level of long-run marginal cost was roughly equal to the estimated 1983 average revenue per kilowatt-hour. However, to contain costs at this level depended on a number of factors. The first was that new, low-cost power generation would, in fact, become available in the early 1990s. The second was that the relatively high growth demand projected for the 1990s would actually occur. Without one of these two conditions, average long-run incremental costs could have been substantially higher.

While the long-run outlook for cost reductions was promising, the utility faced rather severe problems for the remainder of the 1980s. This was the result of the high level of ongoing investments and a temporary, low rate of growth in demand. Because of these two factors, the average incremental costs between 1983 and 1991 were estimated to be much higher, almost US$0.304/kWh. This was close to four times higher than the average 1983 revenue per kilowatt hour. Average costs for that time span were projected to be more than twice as high as expected revenues based on existing tariffs.[30] This large discrepancy between economic and financial costs could have significant financial as well as operational repercussions. In situations like this, therefore, long-run marginal cost-based prices must be made subject to a financial feasibility criterion that imposes the condition that tariffs have to be high enough to cover actual cash-flow requirements, including appropriate accumulations for necessary investment expenditures. If such conditions are not imposed, the chronic lack of funds frequently leads to neglect of maintenance, poor operating performance, and unreliable services. The costs of the latter usually are far higher than the financial costs of added tariffs. This has been demonstrated by the data of the case study presented in Annex 4.2.

DEALING WITH INFLATION AND RELATIVE PRICE CHANGES

Throughout this chapter, it has been assumed that prices and costs are expressed in real (i.e., constant) terms. In reality, of course, costs and prices rarely stay put for long but are subject to change instead. To deal with these changes, it is necessary to differentiate between two types: general price level changes, usually called inflation, and relative price changes that affect only specific inputs or resources and change their costs relative to all other goods and services. Relative price changes are "real" price changes in the sense that they usually reflect fundamental changes in underlying, relative cost relationships. The best-known example for such real price changes is, of course, the drastic change in world petroleum prices in the 1970s.

Dealing with inflationary price changes appears simple, at least conceptually. As the general price level increases, it can be argued that the prices of all energy resources should increase at the same rate, based on some form of appropriate price index, such as the wholesale or consumer index. If the markets for a specific energy resource are free and unregulated, most energy prices will tend to follow these trends. However, apart from woodfuels, the prices for most energy products are subject to some form of price controls in almost all countries. Therefore, specific policy decisions are needed to bring about appropriate price adjustments.

In situations in which all other costs and prices are free to rise, indexing energy prices to general price level changes is appropriate. However, if significant sections of the economy are subject to price controls (as, for example, wages and incomes), then the general indexing of energy prices is no longer warranted. In such situations all the underlying cost components of the specific energy supply system have to be identified, and specific indexes should be used to change only those components whose costs are actually changing. These should be changed at rates that are equal to the cost changes of those inputs. The use of such disaggregated price indexes is quite common in long-term, energy contract price negotiations. For example, they form an integral part of the natural gas pricing formula applicable to the supply of Union Pacific gas to Thailand's Gas Authority. Certain percentages of the total composite gas price charged are based on Thailand's domestic price index, while others are tied to US manufacturing cost indexes and yet another proportion to world oil prices.[31]

For revaluing the costs of recurring input expenditures, the above rules are straightforward. However, allowances also have to be made for sunk capital expenditures. For these, a revalued asset approach, based on a revaluation of current replacement costs, is generally the most sensible one. The main purpose for using revalued assets as a base is to protect the real value of the invested capital. If historical costs were used instead, this base would shrink and could be dissipated completely in situations of very high inflation such as those prevailing in most of South America, for example. However, because part of that base has already been recovered through past depreciation allowances, the revaluation of assets should only be applied to the nondepreciated portion of the original assets. If it were applied to the whole, unjustified windfall gains to the assets owner would result.

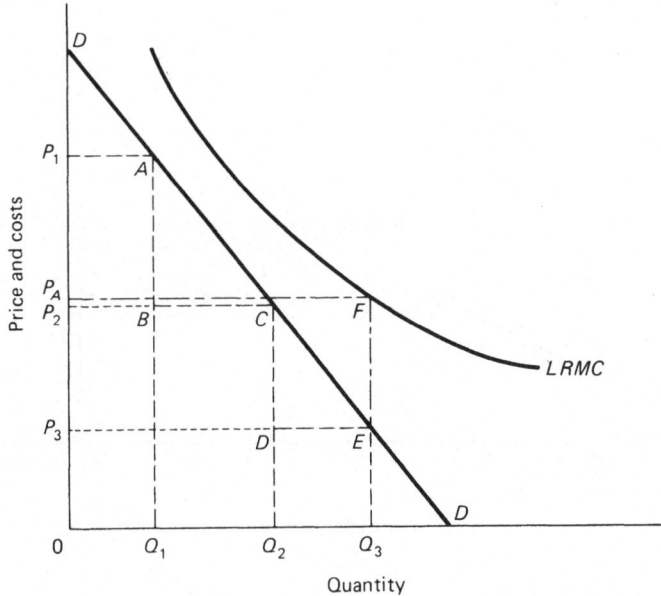

Figure 4.1 Economically efficient price discrimination under economies of scale

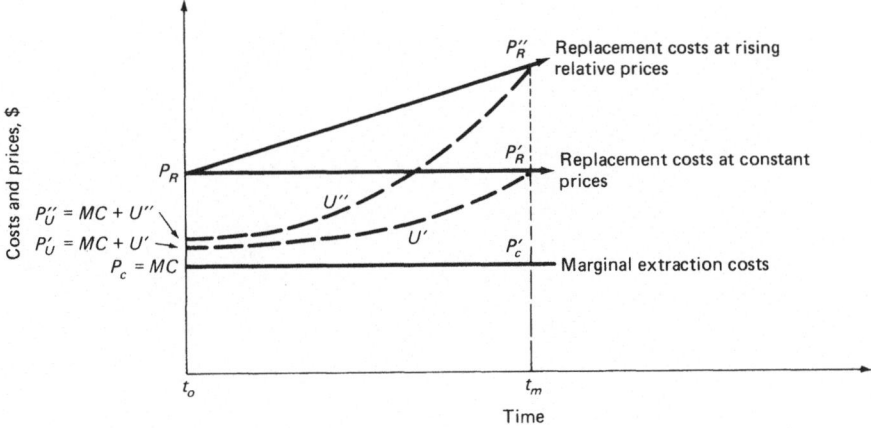

Figure 4.2 Depletion costs with constant and rising replacement costs

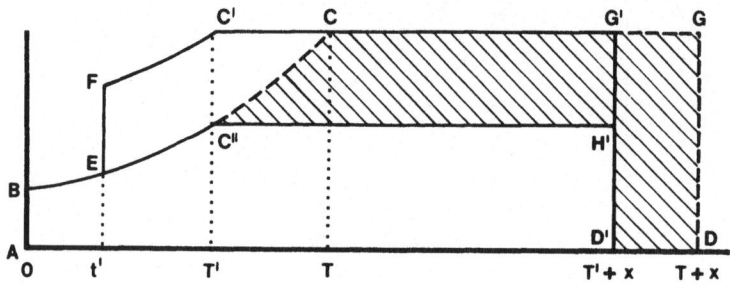

Figure 4.3 Stylized production profile with and without large export project

Table 4.1 The sensitivity of depletion costs to changes in interest rates and reserve/production ratios

Real interest rate (percentage)	Unconstrained life expectancy of deposit (years)[a]	Present value of depletion cost (per US dollar)[b]
4	10	0.50
4	20	0.34
4	30	0.23
8	10	0.26
8	20	0.12
8	30	0.06
12	10	0.15
12	20	0.05
12	30	0.02

[a] Number of years until the reserve/production ratio reaches 15, assumed to be the maximum permissible rate of production; thereafter, rate assumed to be constant for 15 years until exhaustion.
[b] Per US dollar of future net-cost differential between the marginal supply cost of the gas and its next best substitute.

Table 4.2 Representative example of user costs attributable to a large gas consumer (interest rate 12 percent per annum)

Proved gas reserves of deposit	700 billion ft^3
Requested contract amount for single, industrial use	500 billion ft^3

Potential domestic gas demand profile (high forecast):

Year	Without industrial user[a] (MMCFD)[b]	With industrial user[a] (MMCFD)[b]
1	4	4
10	12	12 (maximum)
20	35 (maximum)	exhaustion
32	exhaustion	

Domestic gas supply systems costs:

Present value of total life-time costs	US$99 million	US$54 million
Average levelized life-time supply costs per 1,000 ft^3	1.04	0.95
Average levelized life-time depletion costs per 1,000 ft^3	0.17	1.02
Total costs	1.21	1.97
Depletion costs as a percent of total costs	14%	52%

Industrial plant gas supply costs:

Average levelized life-time gas supply costs per 1,000 ft^3	0.26
Average levelized depletion costs assignable to industrial plant	1.19
Total economic costs	1.45
Depletion costs as a percent of total costs	82%

[a] Based on projected 40-year production profile at an average plant utilization factor of 75 percent.
[b] Million cubic feet per day.

ANNEX 4.1
THE ECONOMIC OPPORTUNITY COSTS OF DOMESTIC CRUDE CONSUMPTION FOR AN OIL EXPORTING COUNTRY SUBJECT TO EXPORT QUOTAS

Total exports:	9.2 million metric tons
Realized value:	US$2,521 million
Average price per ton:	US$274
Sales for domestic consumption:	2.3 million metric tons
Realized value:	US$246 million
Average price per ton:	US$106
Apparent economic subsidy per ton:	US$167

The "apparent" economic subsidy of US$167 per ton would represent the real economic subsidy only if the country's crude could be freely sold at the world market. This was not the case owing to the Organization of the Petroleum Exporting Countries' output restrictions. If these were to be maintained for longer periods, the real economic value of crude used domestically would be represented by the discounted value of future sales. For illustrative purposes these discounted values have been calculated for periods of two, five, and ten years, on the assumption that existing output restrictions may be lifted then.

Interest rate	10%
Projected average rate of increase in real world market prices of crude:	2% per annum
Crude base price:	US$274/ton

Duration of output restrictions	Today's net value per ton
2 years	US$236
5 years	US$188
10 years	US$129

It should be noted that the present net value equivalent per ton would increase each year at the rate of interest (i.e., 10 percent), provided the date on which export restrictions were expected to be removed remained unchanged. The analysis assumes further that output could be expanded almost instantaneously to sell all accumulated oil in the ground. If production rates of "left-over" oil had to be carried out at normal production rates, the opportunity costs would be lower still.[32] With a decline rate of 10 percent, for example, the apparent economic subsidy calculated above would be reduced to zero by an export ceiling of only about three-and-a-half years. However, the conclusions apply only if OPEC quotas exempted crude production for domestic use. If output restrictions were applied to total crude production instead, then the opportunity cost of domestic petroleum product consumption is equal to the full, prevailing f.o.b. world market price, because the oil not consumed domestically could be freely sold abroad.

ANNEX 4.2
THE ECONOMIC COSTS OF UNRELIABLE POWER SUPPLIES

The reliability of electricity services supplied by a public utility in a particular developing country is extremely poor and has reached crisis proportions, with available capacity in early 1983 reduced from a nominal capacity of about 2,800 megawatts (MW) to about 1,200 MW. Forced outages, service curtailments, and damaging voltage fluctuations are daily occurrences; they are the results of inadequate availability of generating capacity and frequent failures of transmission and distribution facilities. At generation and transmission levels above 33 kilovolts (kV), for example, 114 separate individual component faults were reported for November 1981. On 23 occasions between 5 to 27 April 1982, the total system collapsed; 12 of these were related to line faults and 11 to sudden loss of generation. In the summer of 1982, the average plant availability factor was only 53 percent, while the ratio of average weekly power output to installed capacity was only 38 percent. These unreliable power services have resulted in substantial added costs to consumers and economic sectors. The utility itself has suffered because frequent emergency repair work has added substantially to operating costs; it made it necessary to withdraw scarce technical personnel from their ordinary tasks, resulting in a further deterioration of normal operating and maintenance work. To power users, it has created substantial costs by interrupting scheduled activities, reducing output, and greatly increasing the cost of production. To control these costs it has forced users to acquire and operate back-up facilities. The economy as a whole has suffered from the higher costs of production associated with reduced output and productivity, the need to import stand by equipment, and the consumption of valuable petroleum fuels. Basically, however, based on tariff levels only, overall costs of electricity to users represent a relatively modest percentage of total production costs. Using 1980 data as a base, electricity supplies to industry cost US$178.6 million, or about 3.5 percent of the total value of industrial output. For the commercial and service sectors, electricity costs were US$102 million, or 0.7 percent of the value of sales and services. Sales to the residential sector were about US$235.9 million.

However, the costs of electricity not supplied due to outages has raised the total imputable costs of electricity by a substantial margin. Unfortunately, no country-specific data were available to give an indication of the actual losses suffered by users from both outages and voltage fluctuations. However, data gathered elsewhere in the world are indicative of such costs. In Brazil, for example, the net costs of outages to industrial power users from reductions in value added were found to be US$2–10/kWh of electricity not supplied (converted to 1982 dollars). In the residential sector, it was found that the expressed willingness of users to pay to avoid service interruptions per hour was roughly equal to the hourly wage rate.[33] If, for illustrative purposes, it is assumed that average industrial outage costs in the country amount to US$3/kWh not supplied and to US$1.50 for residential and commercial users,[34] then total economic losses in 1980 could have been US$1 billion to industry (at an average effective outage rate of 20 percent) and to US$0.7 billion to commercial and residential sectors (at an average effective outage rate of 15 percent).[35] These losses are far in excess of the revenue received for the electricity supplied by the utility, which amounted to about US$518 million for 1980.

Most users of electricity are quite aware of the high costs of supply interruptions, and those who are financially able to do so have protected themselves by installing stand-by generating equipment. Use of such stand-by equipment is widespread in all sectors of the economy, and it can be assumed that stand-by units back up most, if not all, of the larger industrial/commercial loads. These, in 1982, accounted for a peak demand of some 600 MW. It is estimated that at least 600 MW of stand-by equipment is in regular use. Allowing for diversity between consumers and a margin of capacity, the installed stand-by capacity more likely is around 1,200–1,500 MW.

The additional costs of owning and operating this equipment have been estimated at US$183/kW of capacity a year. If stand-by equipment were not needed and the utility could reliably supply all power needs, the cost of electricity provided by the utility to larger industrial users would be US$503 per kilowatt-year. The added financial costs of using company-owned stand-by equipment at current diesel fuel prices of US$4.75 million Btu would have been US$79 per kilowatt-year of demand. This is an apparent increase of only 16 percent, a reflection of the relatively high tariffs charged by the utility. However, if diesel fuel prices would have been raised to reflect economic costs of US$8.63 million Btu, the differential costs would have amounted to US$152 per kilowatt-year, or about 30 percent more than the cost of utility supply alone. Nevertheless, these added costs still appeared rather modest compared with the potential costs of nonsupply from outages estimated above. However, the economic costs to the country as a whole were quite substantial. It was calculated that the economic costs of using stand-by equipment in 1981 amounted to about US$160 million.[36] These costs consisted of the sum of equipment, operating, and diesel fuel costs. If prevailing trends continued and the ratio of stand-by equipment to total industrial/commercial power demand remained the same, these economic costs were projected to rise to US$320 million by 1985, and to US$560 million by 1990 (in constant prices). The relevant economic question, therefore, is whether or not for the same amounts per year the utility's operating performance could not have been improved sufficiently to eliminate the need for most of these stand-by units. For example, the total five-year expenditure on stand-by equipment between 1983 to 1987 had been estimated to be US$2 billion. If a more reliable power supply could eliminate the use of, say, two-thirds of this equipment, expenditures of about US$1,325 million could be justified in improving reliability levels over the next five years. Actual benefits would be far higher than indicated here, because of the continued rapid growth of the system beyond the assumed five years and the high costs of power outages to customers not protected by stand-by equipment.

NOTES

[1]See, for example, Turvey (1968); Turvey and Anderson (1977); Munasinghe and Warford (1982); Webb and Ricketts (1980), Chapter 4; Munasinghe and Schramm (1983), Chapters 4, 5, 9, and 10; Albouy (1983).

[2]Unless there are readily available substitutes at lower prices available to the user, as, for example, No. 6 fuel oil as a substitute for natural gas (or vice versa) as a boiler fuel for installations equipped to use either.

[3]For a detailed evaluation of the economic effects of such taxes, see Chapter 3, by Newbery, in this volume.

[4]The latter two objectives represent true externalities that should always be accounted for in the calculation of marginal costs. However, the difficulties of converting these physical effects into equivalent economic values are well known and may require the setting of physical objectives or standards instead.

[5]For a more detailed discussion of the importance of shadow pricing, see Chapter 6, by Siddayao, in this volume.

[6]In economic jargon this says that "consumer surplus" is present, i.e., the willingness of consumers to pay is higher than the (opportunity-cost-based) prices charged.

[7]A typical example would be a chemical feedstock plant that uses the resource as a feedstock (for which no substitute might be available); as an energy source for power requirements (which could be replaced by electricity from a public supply source); and as a source for general purpose warm water supplies, heating, air conditioning, or other lower priority value uses for which other substitutes are available.

[8]This issue is analysed below.

[9]See also Chapter 3, by Newbery, in this volume.

[10]See Newbery, Chapter 3 in this volume.

[11]The following discussion is abstracted from Munasinghe and Schramm (1983), Chapter 4.

[12]See Chapter 3, by Newbery, in this book.

[13]For a more detailed review of this issue, see Munasinghe and Schramm (1983), p. 156 ff.

[14]In the Thai case, both imported coal and lignite developments, in addition to gas, are subject to scale economies and both domestic gas and lignite are subject to rising marginal depletion costs as output increases. This complicates the optimal choice. See Munasinghe and Schramm (1983), Chapter 10.

[15]An example would be a gas-based fertilizer plant with a high value, but limited domestic, and a low-value export market.

[16]Evaluated at an interest rate of 12 percent.

[17]For a detailed technical discussion of appropriate methodologies to calculate marginal depletion costs, see Schramm (forthcoming); for estimating marginal depletion costs, see Munasinghe and Schramm (1983), Chapter 11.

[18]For an analysis of optimal strategies for the utilization of natural gas resources in developing countries, see Schramm (1984).

[19]The discussion here implicitly assumes that the resource owner is a national government for which total economic benefits or costs are those that accrue to the country as a whole.

[20]To simplify the exposition, it is assumed that production rates would not taper off but remain constant until final exhaustion of the reservoir.

[21]With US$1.19/1,000 ft^3 chargeable to the project, the depletion costs applicable to domestic uses would have remained at the pre project level of US$0.17/1,000 ft^3.

[22]This covered only the minimum long-run marginal economic costs of the gas, excluding the sunk costs of finding and developing the gas deposit. Average financial costs, therefore, were considerably higher. On the other hand, the preceding analysis does not include any net benefits to the national economy from the additional domestic value added of the proposed chemical plant.

[23]For a representative calculation of these net opportunity costs under export quotas, see Annex 4.1.

[24]Which may be crude oil of specific characteristics designed to meet refinery process configurations, or crudes spiked with various proportions of refined or semi refined products to meet domestic demand profiles.

[25]This is the case in Nigeria, for example, where large quantities of LPG are flared or used as refinery fuel. Its ex-refinery prices in 1983 were a low US$79.77/ton, compared to a world market price of US$255/ton. The reason for this discrepancy was a highly inefficient and costly domestic marketing system which made LPG a high-cost product to end users. Also, a lack of appropriate LPG loading facilities did not permit exports at higher prices. Such a lack of LPG handling and marketing facilities is not uncommon in developing countries.

[26]Typically, ocean freight rates of crude, shipped in large-sized "dirty" tankers, are only about 20 to 30 percent of the freight charges for "white" products. This cost differential provides a protective umbrella for domestic refinery operations.

[27]In some countries the rate may be as high as 12 percent, although in others with few or little investment opportunities, it could be closer to 8 percent.

[28]The unit is actually far too large for the power system and thus not optimal economically.

[29]The power of cash-heavy, semi-autonomous governmental organizations is well enough known in many developing countries. Petrobras in Brazil and Pemex in Mexico are prime examples.

[30]Average costs were lower than average incremental costs because the former were based on total, not incremental, systems costs and revenues.

[31]The latter part of the overall price indexation is not a cost-tracking one but reflects changes in the opportunity costs of alternative fuels instead.

[32]I am grateful to David Hughart who suggested this point.

[33]See Munasinghe (1979), Table D7. A recent survey article reports estimated industrial outage costs to range between US$1.20/kWh in Sweden to US$6.76/kWh in urban areas in Finland, with data for the United Kingdom, the United States, Canada, Chile, and others ranging between these two values. Residential outage costs were generally estimated to be lower, ranging from only US$0.05/kWh in Florida to several dollars elsewhere. From Sanghvi (1982).

[34]These rates are less than half of those estimated for Brazil.

[35]The latter were assumed to be lower because of the lower load factors of these consumer categories.

[36]These estimates excluded the apparently rather widespread use of small units by private households, stores, etc.

REFERENCES

Albouy, Y. (1983). *Marginal Cost Analysis and Pricing of Water and Electric Power.* Washington, D.C.: Inter-American Development Bank.

Munasinghe, M. (1979). *The Economics of Power System Reliability and Planning.* Baltimore: The Johns Hopkins University Press.

Munasinghe, M. and G. Schramm (1983). *Energy Economics, Demand Management and Conservation Policies.* New York: Van Nostrand Reinhold.

Munasinghe, M. and J. J. Warford (1982). *Electricity Pricing, Theory and Case Studies.* A World Bank publication. Baltimore: The Johns Hopkins University Press.

Sanghvi, A. P. (1982). "Economic costs of electricity supply interruptions." *Energy Economics,* Vol. 4, No. 3 (July), pp. 180–198.

Schramm, G. (1984). "The changing world of natural gas utilization." *Natural Resources Journal,* Vol. 24 (Spring), pp. 405–436.

Schramm, G. (forthcoming). *Practical Approaches for Estimating Resource Depletion Costs.* Energy Department Paper. Washington, D.C.: The World Bank.

Turvey, R. (1968). *Optimal Pricing and Investment in Electricity Supply*. Cambridge, Mass.: The MIT Press.

Turvey, R. and D. Anderson (1977). *Electricity Economics*. A World Bank publication. Baltimore: The Johns Hopkins University Press.

Webb, M. G. and M. J. Ricketts (1980). *The Economics of Energy*. New York: John Wiley and Sons.

Chapter 5

ENERGY PRICING IN DEVELOPING COUNTRIES: ROLE OF PRICES IN INVESTMENT ALLOCATION AND CONSUMER CHOICES

Ramesh Bhatia

INTRODUCTION

As noted in preceding chapters, the basic objectives of energy pricing policy will include the achievement of (1) economic efficiency and (2) social equity, while maintaining (3) financial viability. Some of these objectives may be translated in terms of criteria of fixing administered prices, e.g., "lifeline rates" for electricity, subsidized kerosene for meeting basic needs, control of inflation, encouragement to domestic resources, optimum investments in fuel-producing sectors, optimum product mix of refineries, and profitability and efficient management of public sector units. Administered prices could be changed at one or more of the following stages: (1) resource pricing, (2) transfer pricing (to conversion units), (3) output pricing (for the manufacturing unit), and (4) consumer pricing through taxes and subsidies. It is important to realize that some of the objectives of energy pricing can be achieved by adjusting the final product prices through appropriate taxes and subsidies. Adjustment of consumer prices is sufficient to attain objectives such as meeting basic needs, controlling inflation, and considering environmental requirements. Thus, distortions in the mine-mouth or well-head pricing of energy resources, transfer pricing, and product prices can be avoided if it is kept in mind that these prices do not have to be used for achieving macro-economic objectives. Thus, for example, the optimum investment pattern in fuel-producing sectors and the optimum mix of refineries can be ensured by setting output prices which reflect their opportunity costs or replacement costs without consideration of their impact on consumers. In fact, a great deal of confusion in the arguments on energy pricing can be avoided if a clear distinction is made between crude oil/gas prices used in the evaluation of benefits from an investment project and prices (transfer) to be charged to the user industries/units. In the same way, relative product prices faced by the refineries can be adjusted to provide incentives for producing the desired product mix (e.g., maximizing kerosene and diesels at the cost of furnace oil) rather than be constrained by the level of consumer prices.[1]

However, in the actual practice of fixing fuel prices, the interrelations between prices at different stages and those among different fuels may get ignored. In order to meet a certain objective, the government may be fixing prices which may not be remunerative for the producers. Under these circumstances, the producing units will not make adequate profits (or may in fact incur losses) for investment in new facilities/modernization schemes. This would result in shortages of supplies, producing adverse impacts on economic and social development. If resources are diverted from other sectors to this particular energy subsector, this diversion would also adversely affect the growth process. The administered prices of fuels and electricity may be such that these do not reflect their opportunity costs. These prices may, in turn, distort the consumer choices in different end-uses such as cooking, lighting, and irrigation pumping. In many cases, the energy prices may be so administered that they may encourage misallocation of resources without meeting the equity objective. Under these circumstances, it becomes necessary to consider policy alternatives which reconcile the objectives of equity and efficiency in the context of energy prices. The purpose of this paper is to discuss the role of energy prices in meeting various objectives and the consequences of taxes and subsidies on energy inputs.

COMPONENTS OF AN INTEGRATED FRAMEWORK FOR ENERGY PRICING

One of the important aspects of energy pricing in developing countries is that prices of different fuels at different stages should be analysed in an integrated framework. Figure 5.1 outlines the scope of such an integrated framework in which links among energy subsectors and different stages of pricing have been shown. The *first stage* relates to pricing of energy resources such as coal, crude oil/natural gas, hydropower, and renewables for evaluating investment options and financial planning at the upstream stage of the energy industry. The important question here is whether the prices used for valuation of output of different energy projects should be based on the opportunity cost of importing/exporting that output (or its substitutes). The *second stage* involves "transfer pricing" issues; these are concerned with the setting of prices to be paid by the units which purchase these energy resources for further processing, e.g., the price of crude oil paid by refineries or price of coal paid by electricity generation firms.

The main issue raised at this stage is how different organizations share the "economic rent" arising from the use of exhaustible resources. The issues are further complicated by the fact that some of these outputs (e.g., natural gas, coal) are also sold to consumers other than energy units (e.g., fertilizer plants, households, industries). The *third stage* relates to downstream producers' prices, i.e., prices which are received by refineries and electric utilities for their output. The important question here is the role of petroleum products price differentials faced by a refinery in determining its outputmix. The issues relating to relative profitability of different units in the energy sector may also be discussed at this stage along with transfer pricing. The *final stage* relates to prices paid by the consumers of final outputs (e.g., kerosene, diesel, electricity, soft coke, charcoal). The important aspects of taxes and subsidies for different energy products are discussed at this stage.

The above discussion of various stages is for the purpose of analytical convenience and may not be appropriate in countries where all these subsectors/stages are not important. Besides, the nature of organizations involved in various stages may differ for each fuel subsector: a few refineries (perhaps government owned) for oil products against a large number of private units manufacturing charcoal/soft coke.

Still, it is important to analyse the relevant pricing issues in the context of an integrated framework as outlined above because:

1. The substitution possibilities differ from one fuel to the other and the elasticities of response to changes in prices of each fuel also vary.

2. The price of a given fuel to the consumer can be varied by changing the prices at one or all of the stages mentioned above. The impact of higher import prices of crude oil can be passed on to the consumers either by increasing transfer prices to refineries or by changing the excise duties on oil products or both. Since different options have different implications for government revenues and profitability of energy units, it may be necessary to analyse the detailed impacts.

3. A given objective of energy pricing policy may not require changes at all the stages. For example, keeping prices of a few or all oil products at levels lower than import costs could be done by fixing low or differential excise duties and does not require fixing a low transfer price of crude oil.[2] Although there have been studies on pricing of individual fuels, oil products, and electricity, there are very few studies which analyse pricing issues in an integrated manner.

PRODUCT PRICES, OUTPUT, AND INVESTMENT

One of the important objectives of pricing policy is to raise adequate revenues to meet operating expenses and make provisions for investment. In many developing countries energy-producing enterprises are in the government sector, although some private sector units may operate in some areas. In the case of public sector units, raising revenues is not given as much importance as it deserves, and it is sometimes argued that public sector units should not be concerned with making profits since they have to fulfil other social objectives. As a result of this feeling about the role of the public sector, large-scale financial losses in these units are tolerated. It is also argued that the financial profits or losses should not be the criterion for judging public sector units because funds for investment can always be transferred from the general resources pool. However, it may be mentioned that losses in public sector units may be due to various factors including low output prices, high wage costs, high capital costs, and management inefficiencies. Hence, it would be necessary to isolate the effect of output prices from other factors so that measures to achieve improvements in efficiency can be carried out.

Given the cost structure and management efficiency, low product prices would result in low sales realization, lower profits (or losses), lower retained earnings, and lower investments. Thus, unremunerative prices for output would result in losses (or low profits) for the undertaking which, in turn, would affect its ability to finance new investments. Since transfer of resources from other sectors to the energy sector would result in slowing down the progress in other sectors/programmes, the inevitable result of low prices is inadequate investment in the energy sector/subsector. Lack of investment in production and distribution of energy leads to shortages which, in turn, affect economic and social developments in the country. In this section we present a few examples of how low product prices affect allocation of energy investments and production. The illustrations are usually given from India since data at the required level of disaggregation were not available for other countries. It is hoped that generation of similar data and analyses for other countries would be one of the outcomes of the Energy Pricing Policy Workshop.

Output prices, production, and investment in the coal sector of India

India is the world's seventh largest coal producer, with a coal production of 130 million tons in 1982–83. Coal provides nearly one-third of India's commercial energy (in coal equivalents) and is expected to play a critical role in providing a large share of commercial energy supplies for the next several decades. India's coal reserves are large — more than 80 billion tons of reserves and resources compared with about 6.5 billion tons of prognostic recoverable hydrocarbon resources.[3]

Following nationalization, coal production increased rapidly from 78.1 million tons in 1973–74 to 99.7 million tons in 1975–76, at which point production was adequate to meet demand. Subsequently, however, production stagnated and fell behind demand, reaching only 104 million tons in 1979–80. It has since increased to 114 million tons in 1980–81 and to 130 million tons in 1982–83.

During the periods 1975–76 to 1978–79, the wholesale price index of coal changed from 146.9 in 1975–76 to 211.5 in 1978–79 (1970–71 = 100). Coal prices are set by the government on an administered basis and are intended to at least cover production costs, with capital expenditure funds being provided from central public investment funds. In practice, prices have not been sufficient to cover operating costs and, as a result, the coal companies have been dependent on the government not only for capital expenditure funds but also for financing cash losses.

The situation of low output prices leading to financial losses was particularly serious from 1975 to 1979. In 1974, the Fernandes Committee recommended prices based on average cost of production for the whole sector, excluding interest on debt, but including a 10 percent return on capital. The government accepted the recommended price increase in April 1974, but rejected the proposed 10 percent return on capital from the average cost estimate. The Chakravarty Committee price recommendations (May 1975) were based on estimated average cost of production in 1975–76, excluding interest on short-term, non-Plan loans (i.e., loans to cover losses), but allowing for return of 5 percent on equity. However, price revisions announced by the government in July 1975 did not allow for return on equity as well as depreciation.[4]

Output prices and profitability. As a result of low prices, average sales realization for coal[5] in India has been lower than total costs in all the five years from 1976 to 1980. In fact, for the first three years, average sales realization did not cover even the operating costs which increased substantially due to high wage costs. The average financial loss in 1978–79 was Rs 28.7 per ton and the total losses in that year were on the order of Rs 2,379 million (approx. US$250 million). The total losses for the 5-year period were Rs 5,839 million (US$600 million).

Due to the large operating losses during these five years, the government of India had to provide funds to cover financial charges and losses. In the five years from 1976 to 1981, the government provided Rs 16.8 billion to Coal India Limited (CIL), of which Rs 4.6 billion was non-Plan support to cover the financial gap due to previous losses. The Plan support of Rs 11.7 billion almost exactly matched the capital expenditures for the period. This analysis shows that if the coal prices were raised during the period 1976 to 1979 to cover increases in costs, an additional amount of almost Rs 5 billion would have been available to meet the investment requirements of the coal industry. This would have increased the total investments in this sector by more than 45 percent.

A comparison of profitability of the coal industry vis-à-vis other energy subsectors also shows that the financial performance of coal was worse than that of the other two sectors — petroleum and electricity. According to the report of the Working Group on Energy Policy (Government of India, 1979), "The petroleum sector has a rate of profit (gross rate of return, i.e., gross profit as percent of total capital) which is substantially higher (27.4 percent) than for the public sector as a whole (8.6 percent); the electricity sector is slightly below the average (8.5 percent) and the coal sector is considerably below it (0.6 percent)." The report suggested that the basic objective of the energy pricing policy should be to (1) generate sufficient surpluses to facilitate the financing of investments in the energy sector, (2) induce economies in the use of energy in all sectors, and (3) encourage the desired forms of inter-fuel substitution. The report pointed out that the then pricing policies in India did not subserve the first two objectives, while the third objective was being met only partially.

Product prices and opportunity costs. Another aspect of coal pricing in India is that not only have the coal prices been set below the average cost of production,[6] but they have been substantially lower than the international prices. In the early 1980s, f.o.b. steam coal export prices were in the range of US$50–57 per ton. Taking into account variances in f.o.b. prices due to size and duration of contracts, a reasonable export price[7] for Indian coal (for coal with 6,200 kcal/kg, 0.6 percent sulphur, and 16 percent ash) was estimated by the World Bank as US$50 per ton. This would be comparable to a minehead price of about US$45 per ton, after adjustments for transportation and handling costs. By comparison, such coal was priced at US$18.5 per ton (Rs 168/ton). On average noncoking coal in India has a calorific value of 4,500–5,000 kilocalories (kcal)/kilogram (kg) which, after adjustment of heating value and quality differences, would indicate that the average economic value for noncoking coals was in the range of US$30–35 per ton (in 1981–82) based on international prices.[8] Thus, the market price of coal in India was fixed at about one half of the border price and, hence, did not reflect its opportunity cost to the consumer.

Given the past and current shortages of coal, one may consider the shadow price of coal as equivalent to the shadow price (or opportunity cost) of an alternative fuel, i.e., fuel oil. When corrected for the differences in heat content and efficiency of use (two tons of coal equal to one ton of fuel oil),[9] the cost of coal for steam raising was approximately US$37 at market prices compared with US$194 for a ton of imported fuel oil.

One may argue that coal prices should not be set at parity with imported fuel oil because of the following:[10] (1) India is pursuing a policy of encouraging substitution of coal for hydrocarbon fuels, about half of which are imported, imposing a heavy burden on balance-of-payments.[11] A premium may be attached to diversification away from hydrocarbon fuels where import dependence is high and stability of supply sources may be uncertain. (2) The price differential between coal and imported fuel oil has been extremely large, and the inflationary impact and the consequent social costs of basing prices on full opportunity cost would be unjustifiably high. (3) Indian society (consumers) should be allowed to benefit from the use of an indigenous resource that is not priced artificially high.[12]

Impact of low product prices. The question of pricing of coal in India brings into focus the following issues: (1) What should be the relationship, if any, between the price of a domestic fuel and its opportunity cost? (2) Should there not be differential prices to be paid to the producers, to be charged to the consumers, and to be used for evaluating investment decisions? How should these prices be determined?

One may accept the reasons given above for not fixing coal prices at parity with fuel oil, but there is no justification for fixing prices which do not even cover the costs of production (sometimes, not even the operating costs are covered). The spread between the opportunity cost of coal and domestic prices charged to the consumer is too large — approximately US$37 for an equivalent quantity of coal as against US$194 for imported fuel oil. This shows that the consumers are getting coal at one-fifth of the price at which an alternative fuel would be available. The result of these subsidies on coal production (and also on transportation) is that there are no incentives to improve efficiency in the use of coal. The "economic rent" on the domestic resource is being totally passed on to the consumers of coal, and the society is not recovering a part of it to be spent on developing resources and improving the environment. Since low prices had led to stagnation in output of coal, the shortages of coal were met by increased imports of fuel oil. During the periods 1975–76 to 1981–82, India imported 5.56 million tons of furnace oil involving a foreign exchange outflow of US$605 million.[13] Hence, it becomes necessary to take a rational, integrated view of the situation in fixing coal prices. By avoiding the inflationary effects[14] of increase in coal prices, the country has burdened itself with imports of fuel oil which have adverse effects on the economy, both direct (import costs are higher) as well as indirect (reduced imports of other commodities).

Prices used for evaluating investments. A rational coal pricing policy in India would be to make a clear distinction between the price to be used in estimation of benefits in an investment project and the prices to be charged to the consumer. In the Indian context, in the medium term, additional quantities of coal would substitute for imported fuel oil or would release fuel oil for exports. In this situation, the coal price to be used to measure the economic benefits of proposed mine developments should be the border price (f.o.b. or c.i.f.) of fuel oil adjusted for the shadow price of foreign exchange. In the relatively longer period (more than 10–15 years), the coal price for evaluating investments should be the f.o.b. export value of coal, less internal transportation and handling costs. Of course, this is true for India because India can, in fact, export high grade coal if it is shown to be economic to do so.[15] For some other countries where coal exports would not actually take place (e.g., for Thailand and the Philippines),[16] it would not be correct to link coal prices with coal exports. However, even in these countries, if additional availability of domestic coal (or natural gas) reduces oil imports at the margin, the relevant price in evaluating investments in coal mining would be the c.i.f. price of oil product (e.g., fuel oil) adjusted for the shadow price of foreign exchange.

The significance of using appropriate prices for evaluation of benefits in investment projects can be illustrated by giving data on allocation of investments in the energy subsectors in India. In the Sixth Five Year Plan (1980–85), the total outlays on the energy sector were as follows (in 1979––80 prices, approximate U.S. dollars[17]): US$19.265 billion for electricity, US$2.87 billion for coal, and US$4.3 billion for petroleum. In July 1982, additional funds on the order of US$2.5 billion were sanctioned for the petroleum sector, essentially for crude oil production. Even though the coal sector[18] has used up its total outlays in the first three years, no additional funds have been allocated to the coal sector. Although these figures do not provide conclusive evidence of the method of allocation of investments in various subsectors within the energy sector, one cannot rule out the impact of "high" profitability of investments in the oil sector as against the "low" profitability of investments in the coal sector on allocation of funds.

Prices for the producers. Even though allocation of investment is made on the basis of shadow prices, the prices for the producers need not be fixed equal to shadow prices. The producers may get a price which covers operating costs, given certain minimum standards of efficiency, and provide surplus funds sufficient to meet the capital expenditure requirements necessary for replacement purposes to maintain production capacity in existing mines, as well as to invest in new mines. This would amount to fixing the price equal to long-run marginal cost including a "reasonable profit."

Prices for the consumer. The price to be paid by the consumer should lie somewhere between the producer's price and the shadow price. A reasonable price for the producer would be US$25 per ton which covers costs and provides a 10 percent return on capital. However, the corresponding shadow price is as high as US$100 if c.i.f. value of imported fuel oil is taken. It is difficult to suggest the exact price between these two levels since it would depend on the impact of sudden and substantial increases in coal prices, as well as the extent of resource requirements for the coal sector. A recent report[19] submitted to the government of India has suggested that the consumer price should be fixed at approximately US$33 per ton, about US$8 per ton higher than the pithead price for the producer. This increased price for the consumer is expected to mop up the "consumer's surplus" which is being enjoyed by some users of coal. This additional amount (approximately US$800 million) would be placed in a separate Coal Development Fund and can be used for investments in new mines and for modernization of existing mines. This shows that the recommended price is much higher than the 1981–82 price, as well as the current price of coal. The consumer price is almost equal to the shadow price if f.o.b. export value (without premium on foreign exchange) of coal is taken. It is considerably lower than the shadow price if c.i.f. value of imported fuel oil is considered.

Product prices and production of soft coke in India

Soft coke, as produced and marketed in India, is manufactured from coals with some coking properties. It is used for household cooking, as well as for input in brick kilns. Production of soft coke by Coal India Limited has been declining from 3.25 million tons in 1976–77 to 2.41 million tons in 1979–80 and to 1.74 million tons in 1982–83. This lower availability and use of soft coke would have resulted in high consumption of kerosene and/or fuelwood in cities and small towns. This reduction in the output of soft coke has been due to a variety of factors including:

1. Though kerosene is subsidized by the government up to 25 percent of its c.i.f. price,[20] the subsidy on soft coke has been only US$4 per ton.[21] Even the effect of this subsidy is partly eroded by the royalties/cesses attracted by coal (used as input in manufacture of soft coke), mainly from the state governments.

2. At present market prices, soft coke is at a considerable economic disadvantage with respect to kerosene and LPG because its calorific value is about 60 percent of that of kerosene or LPG, and the efficiency in use, or the appliance efficiency in the case of soft coke, is only about 20 to 25 percent compared with 50 percent and above in the case of kerosene and LPG. This means that, although in terms of Rs/kg soft coke is cheaper than kerosene and LPG, in terms of Rs/kcal (Rs/kg divided by kcal/kg) and Rs per effective kcal (Rs/kg divided by kcal/kg multiplied by appliance efficiency) kerosene and LPG are cheaper than soft coke. Even when LPG and kerosene are valued at import parity prices, soft coke has only a marginal advantage, and that too in specific locations in the eastern region. Besides, it is more convenient to use kerosene/LPG devices as these can be turned on and off whenever the consumer requires it, and the flame is of uniform intensity.

3. Transportation cost is a major component in the market price of soft coke. Average transport costs have been increasing over time as the dispatches by rail have declined from 1.4 million tons in 1976–77 to 0.54 million tons in 1982–83.

4. The quality of soft coke has declined over time since there are no differential prices based on quality. If overall profitability is the criterion, the local management in coal mines tends to ignore quality if that helps to improve the price of run-of-mine coal by reducing coal allocation for coke making.

5. Since soft coke has not been given the same level of subsidy as that given to kerosene, the producers have not been getting remunerative prices.

According to a recent estimate, the producer will incur a loss of US$2–10 (on a soft coke price of US$17.5 fixed by the government), depending upon the type of coking coal used as input.

6. The result of low prices by the government has been that soft coke has been produced in a traditional manner[22] without any control on quality. Investment in large modern plants has not been made because of low anticipated demand which, in turn, is due to high consumer prices and low quality of soft coke. For example, the Government of India recently turned down an investment proposal for manufacturing one million tons of soft coke involving a total capital cost of US$24.2 million. This investment was not considered attractive even when the pay-back period was seven years at the existing controlled price of soft coke. The Bureau of Industrial Costs and Prices of the Government of India has recently recommended that urgent consideration be given to the one million ton plant — with an increased subsidy of US$10 per ton for the consumer — so that the market can be flooded with soft coke. With the higher subsidy, the market price of soft coke would come down to US$11 per ton so that more people might be inclined to switch back to soft coke from kerosene (which is imported, involving foreign exchange). Further, a reduction in the price of soft coke would influence consumers of firewood and charcoal not to switch to kerosene as their incomes increase but instead continue to use soft coke because of the price differential. A lower price of soft coke would also stem the flow of firewood logs from rural areas to the cities.

7. A remunerative price of soft coke would also encourage investment in manufacture of high-quality (smokeless) soft coke and towngas by using a low- or medium-temperature carbonization method. This smokeless soft coke would induce people back from the use of kerosene to soft coke.

Thus, the case of soft coke in India illustrates the following aspects of pricing policies in the energy field:

(a) Distortions were introduced in the consumption pattern when the price of one fuel (kerosene) was subsidized, while that of a close substitute (soft coke) was not subsidized (or at least not subsidized to the same extent). This resulted in the shift from soft coke to kerosene involving a reduction in demand and output by 5.6 million tons[23] in a 5-year period from 1977 to 1982. Assuming that this reduction in availability of soft coke was replaced by imported kerosene, this would have resulted in an increase in kerosene imports by 1.56 million tons.[24] The estimated cost of these kerosene imports was on the order of US$450 million over a 5-year period, an expenditure which could be easily avoided. (b) The price paid to the producer was not made remunerative enough to improve the quality of soft coke and to make investments in new units. For example, the price obtained by the producer was not distinguished by grade so that there was no incentive for the producer to maintain the quality of coals used for soft coke. The result was bad quality coals resulting in low-quality soft coke, reducing its demand further. There was also no incentive to modernize the plants used for manufacturing soft coke in order to improve the quality of soft coke and recover tars. Besides, investment in modern units and Low Temperature Combustion plants was not allowed since it was feared that there would not be adequate demand for soft coke. In this way, an unimaginative pricing policy for soft coke resulted in a vicious circle of low demand, lower quality; low investment, low output, and lower supply. This, in turn, resulted in a foreign exchange outflow of US$450 million for kerosene imports over a five-year period.

Electricity pricing, profitability, and investments

Another example where low output prices lead to low profits (or losses), low investments, and supply shortages is the electric power sector in India. As pointed out by the Committee on Power,[25] the State Electricity Boards have been supplying electricity at subsidized rates to agricultural, industrial, and other groups of consumers. The results of this policy of subsidized rates of different consumers are:

1. The revenues of State Electricity Boards are not sufficient to cover costs and, hence, they incur heavy losses. Because of these losses, they are not in a position to invest in additional generating/transmission capacity and in modernization of existing equipment.

2. The losses on account of low tariffs for certain categories of consumers are partly made good by raising tariffs for other consumer groups (e.g., domestic consumers). Such cross-subsidization has resulted in transfer of resources as between different categories of consumers (e.g., agricultural consumers being subsidized by domestic consumers in urban areas). Such redistribution of resources may not be equitable since agricultural consumers may be relatively better off than urban domestic consumers. Besides, the inability of the State Electricity Boards to earn adequate surpluses to meet their commitments has resulted in the requirements of the power sector being largely met from public taxation. This has led again to a transfer of resources from taxpayers to consumers of subsidized electricity.

3. The subsidy on electricity distorts the relative prices faced by the consumers (e.g., farmers) and their choice of technology is affected by these considerations. These choices may not be in line with the ranking of alternatives from the viewpoint of society and may also result in overcapitalization in the consuming sectors. The questions of price distortions for the consumer, as well as equity considerations, are discussed in later sections. This section is devoted to a discussion of electricity prices, losses, and investments in India.

Electricity pricing and investments in India. In India, industry and agriculture are the two largest consumer groups, accounting for 64 percent and 14 percent, respectively, of the total energy sold. In the industrial category, power-intensive industries (such as aluminium, calcium carbide, and fertilizers where power costs form a significant proportion of operating costs) have generally benefited from highly subsidized rates from the electricity boards within whose jurisdiction they are located.[26] In the case of agriculture, the bulk of the power supply (more than 75 percent in most states) goes to the category of "private tubewells/pumpsets." It is well known that agricultural tariffs are much lower than the average cost of supplying power to the rural areas. Apart from low tariffs, there has been an increasing tendency to shift from metered supplies to flat tariffs related to the horsepower of the pumpset used on the grounds of administrative convenience, saving in cost of meters, and overcoming the problem of theft. States like Rajasthan, Bihar, Punjab, Haryana, Uttar Pradesh, and Maharashtra have adopted the system of flat rate tariffs. Recent tariff rates and the merits and demerits of alternative tariff systems are discussed elsewhere. Here attention is devoted to a comparison of agricultural tariffs vis-à-vis the average cost of power supplied to low-tension consumers in rural areas. The Report of the Committee on Power (1980) has given comparative data for agricultural tariffs in April 1979 and shows the gap between these tariffs and the average cost of supplying power to low-tension consumers. The agricultural tariffs varied from 7.4 paise (0.74 US cents)/kWh in Bihar to 1.0 US cent/kWh in Kerala to 3.5 US cents (Rs 0.35)/kWh in West Bengal. These figures refer to the average cost of supplying low-tension power to urban and rural consumers together, because due to lack of data it was not possible to calculate the cost of supplying the two groups separately. As urban consumers represent relatively concentrated load centres, it can be assumed that the real cost of supplying the rural consumers was considerably higher than the figures mentioned above. For example, in the case of Uttar Pradesh, it was estimated that the cost of supplying power to rural areas was 6.7 US cents (Rs 0.67)/kWh as against the average cost of low-tension power estimated at 4.4 US cents (Rs 0.44)/kWh. Significant variation from one state to the other in recovering the costs of supplying power to rural areas was found.

A study[27] of actual tariffs charged and marginal cost-based tariffs for different users of electricity shows that although existing tariffs for all categories of consumers have increased during the period from 1974 to 1977, they are still considerably below marginal costs. Gellerson (1979) states that this discrepancy is greatest for agricultural consumers, who pay the lowest average tariffs and yet for whom electricity is probably most costly to supply. Thus, agricultural consumers are most heavily subsidized by State Electricity Boards. In the southern region, the average revenue realized is only 16 percent of the marginal cost of supply to the agricultural consumers. In other regions as well, the ratios are rather low: 20 percent for northern and eastern regions, and 34 percent for the western region. The ratio in the western region is higher not because the revenue realized is higher but the energy cost at bus bar is estimated to be lower for the western region compared with the others. These data are a good index of the subsidies provided to the rural consumers of electricity, and the private profitability of using electricity in place of alternative sources may be attributed to the artificial price advantage offered to consumers of one form of energy but not to those using other forms of energy, which could be good substitutes for electricity.

The results of low tariffs for electricity charged to the rural consumers can be seen in the estimated losses incurred by the State Electricity Boards. In 1976–77, the rural electrification losses were estimated to be Rs 156.8 crores (US$150 million), and all the State Electricity Boards reported losses ranging from Rs 4.3 crores for West Bengal to Rs 20.8 crores for Tamilnadu, Rs 15 crores for Punjab, and Rs 14.4 crores for Haryana. However, in 1976–77, an aggregate surplus of Rs 45 crores was shown for the State Electricity Boards taken together which was wiped out by the losses on account of rural power supplies. Thus, the net losses in 1976–77 were on the order of Rs 112 crores (US$112 million) in aggregate.

At the time of formulation of the Sixth Five Year Plan (1980–85), the commercial losses of the State Electricity Boards were estimated at Rs 4,400 crores (US$4,400 million) at 1979–80 rates.[28] The Plan had envisaged that by way of improvement of their financial working, the boards would aim at reducing these losses by 80 percent, (i.e., by about Rs 3,500 crores). However, this expectation has not materialized. On the basis of present assessment, these losses are estimated at about Rs 4,300 crores. Although Boards have taken measures to reduce losses, the increases in costs of inputs and reduction in revenue due to shortfalls in generation have wiped out the surpluses from higher tariffs.

It would, of course, not be correct to say that the poor financial condition of most of the State Electricity Boards is due exclusively to irrational tariff structures. Factors such as poor operating efficiencies, escalating project costs owing to poor project planning and management, and increasing transmission and distribution losses all play their own part. However, it has been suggested that even if operating efficiencies and the utilities' management reach reasonable norms of performance, the present tariffs do not, in the case of most State Electricity Boards, cover total costs, let alone bring the Boards a reasonable rate of return on invested capital.[29] The result of continuing financial losses of the Boards is that their contribution from internal resources[30] to investible funds is rather low.

Since diversion of investible resources from other sectors to power has its own socio-economic implications, the investments in the power sector have never been adequate to meet increasing power demands. There have been continuing shortfalls in targets of installed capacity and energy generation. According to the Working Group on Energy Policy, "inspite of manifold increases in generating capacity (since 1950), power shortages have been experienced in various parts of the country during the past several years. The basic reasons for these shortfalls have been the continuous slip-

pages in the achievement of targets of additional generating capacity."[31]
The percentage shortfall was 50 percent over 1969–74 and 18.4 percent
over 1974–79. These continuing shortages have led to serious impacts on
the economy as follows: (1) losses in agricultural and industrial output
due to non-availability of power, (2) losses and damage to equipment due
to voltage fluctuations and higher input use due to frequent shutdowns
of continuous process plants, (3) captive power generation in a number
of industrial units and commercial establishments resulting in higher capital
costs of electricity generation, and (4) overcapitalization in the rural/
agricultural sector on account of unreliable power supply. A detailed discussion of these issues would be outside the scope of this paper. However,
one illustration of the seriousness and magnitude of the problem is given
by discussing the power situation for the agricultural sector.

The impact of power supply on agricultural sector. During the decade of
the 1970s, the total installed generating capacity in the country increased
from 14,709 MW in 1970–71 to 26,680 MW in 1978–79. For the same
period, the agricultural connected load increased from 6,225 MW to 13,850
MW. The year-wide analysis shows that in the first three years, the annual
increment in total generating capacity was lower than the corresponding
increment in agricultural connected load. Even in other years, the
increments have not been substantial enough to meet the incremental
demands from the rural sector. Assuming that at least two thirds of the
incremental capacity would go for industrial, domestic (urban), and
transport sectors, the annual increments were very inadequate to meet the
demands in rural areas.

It may be mentioned that the figures of connected load need not represent the total power demand from the rural sector. Because agriculture
is a seasonal activity, the agricultural loads have a tendency to coincide
at the time which is generally preferred by farmers (e.g., four hours in the
morning). Besides, due to uncertainty of power supply (and rostering of
supply for fixed hours), the consumers have a tendency to switch on their
motors as soon as power is available. This results in bunching of loads
which can be reduced only if the power supply position is improved and
is made reliable.

The effects of shortages and uncertainty of power supplies are seen in
consumer responses which include (1) installing electric motors of sizes
which are bigger than their requirements so that the water pumping work
can be completed in shorter time, and (2) purchase of "back-up" systems
in the form of diesel engines/motors so as to meet requirements when electricity is not available. In the case of farmers these back-up systems may
be diesel pumpsets, while for industrial units a diesel engine may provide
the necessary back-up support.

Although there are no survey data for various regions to give reliable
estimates for the number and cost of these back-up systems, the impression

is that a large number of consumers (commercial establishments, large farmers, industrial units) maintain a diesel engine for use when electricity is not available or when there are serious voltage fluctuations. A survey of 109 farms in 19 districts of Punjab (in North-West India) showed that 40 percent of the farmers owning electric motors also owned a diesel engine. About 30 percent of the farmers owned one electric motor plus one diesel engine, while the remaining owned more than one electric motor or diesel pumpset. Most of these farmers were large farmers with operating holdings of above two hectares. Although ownership of different types of capital equipment by a farmer is a complex matter depending on many socio-economic and infrastructural factors, it cannot be ruled out that a mix of electric motors and diesel engines does reflect the consumer response to shortages of power.

The extent of such a back-up system would depend on the reliability of power supply, the opportunity cost of not providing irrigation or the returns from timely availability of water (which would depend on rainfall, use of high-yielding variety seeds, use of fertilizers, etc.), availability· of traditional modes of irrigation (e.g., animal-powered devices), and availability of tractors/diesel engines for off-farm operations. For example, in Bihar, a rough estimate is that almost half of the large farmers owning an electric pumpset also owned a diesel engine. In a field visit in 1981 to Saharsa town and nearby villages in north Bihar, it was found that all industrial units operating on electricity had back-up diesel engines.

Although it is hazardous to make a guess about the extent of overcapitalization on account of the factors discussed previously, a rough estimate would underline the dimensions of the problem. If it is assumed that about 20 percent of the 4 million owners of electric pumpsets in the country maintain a diesel pumpset[32] mainly as a back-up for irrigation, this would give a total figure of 0.8 million diesel pumpsets for this purpose. This is not an unreasonable proportion of the total stock of diesel pumpsets approaching 3 million units. Assuming that, on average, a diesel engine costs at least Rs 4,000, this would give an estimate of overcapitalization of Rs 320 crores. To this we may add a notional figure[33] of Rs 50 crores for industrial units. Besides these avoidable capital costs which are incurred by a smaller portion of the total consumers, there are a large number of other consumers who tend to use tractors for pumping water which uses diesel oil very inefficiently. Thus, the order of magnitude figures indicate that there is a need for a thorough evaluation and quantification of this aspect of rural electrification investments, power shortages, and tariff policies.

ROLE OF ENERGY PRICES IN CONSUMER CHOICES

Energy prices in many developing countries are administered prices as fixed by the government. On account of various socio-economic objectives, these prices have elements of taxes or subsidies because of which they do not reflect the real resource costs of using these resources. Invariably, the subsidies or taxes are arbitrarily determined, resulting in distortions in consumer choices. A complete discussion of the effect of relative prices on consumer choices would be outside the scope of this paper on account of time and space constraints. However, the main points of the issues concerned are illustrated with the help of two examples from India: (1) petroleum product prices and their effect on consumer choices, and (2) relative prices of electricity and diesel and their effect on choice of technology in irrigation pumping. The section also discusses the role of prices of conventional energy sources in diffusion of technologies using renewable energy sources.

Petroleum product prices and consumer choices

One of the cases in which relative prices of oil products can distort consumer choices relates to retrofitting of a car with diesel engine for use as a taxicar on intercity routes. In India, petrol or motor gasoline has always attracted high excise duties since it is considered an easy method of raising revenues. Besides, it is also considered an effective demonstration of socialist policies of the governments by which an item of consumption of the rich is being heavily taxed. Currently, the excise duty component is Rs 2.2 per litre in a market price of Rs 6.2 per litre, (i.e., 35.5 percent). The excise duty component of other products is much lower: 18 percent for kerosene, 10 percent for high-speed diesel oil, and around 5 to 10 percent for furnace oil. Mainly because of this differential in excise duties, there is a significant difference in the market prices of petrol and high-speed diesel oil. This differential has provided incentives for consumers to shift away from petrol-using vehicles to diesel-using vehicles. An increasing number of diesel cars, motorcycles, jeeps, and minibuses have come on the road in recent years. On intercity and long-distance routes where fuel costs form a significant proportion of total operating costs, minibuses and cars retrofitted with diesel engines are becoming popular.

Given the relative price of petrol and diesel oil, a private taxi operator finds it economical to spend an extra US$2,500 for retrofitting of a diesel engine,[34] thus increasing the capital cost from US$7,000 to US$9,500 (Table 5.1). However, with the retrofitted diesel engine, his annual fuel costs are lower by around US$2,130. Even when the high costs of repair and overhaul are taken into account, there is a saving of US$1,800 or so, indicating that the pay-back period for a retrofitted diesel engine is about one year and four months; in other words, the total annual costs of running a taxi with retrofitted diesel engine (US$4,460) are lower than those for a car with petrol engine (US$5,698) by a margin of US$1,238 (Table 5.1). This significant saving partly explains the increasing popularity of diesel-driven taxis in northern India.

However, this advantage is accruing to the user because petrol prices are artificially kept high through excise duties. This alternative, though financially profitable for the consumer, need not be a preferred option from the viewpoint of society. This has been shown in Table 5.1 where comparison has also been made by taking the shadow prices of diesel and petrol.[35] By using the shadow prices (which assume these to be equal to c.i.f. prices adjusted for 25 percent premium on foreign exchange), it has been shown that the annual costs of running a diesel-retrofitted car are higher (US$4,329) as compared with the petrol-driven car (US$4,125). The social costs of running a car with petrol would be even lower when it is considered that a relative surplus of petrol and naphtha is likely to develop in the future, while diesel oil will always remain an imported product at the margin.

The foregoing analysis shows that by artificially pricing petrol at a high level (or by not being able to raise the diesel prices to the same level), the government has encouraged the shift to diesel vehicles, which is not a desirable shift from the viewpoint of society. The demand for diesel oil has been increasing much faster than for other oil products[36] and has increased by about 2 million tons (from 9.8 to 11.8 million tons per year) in the past four years. Almost 2 to 3 million tons of high-speed diesel have been imported per year in the past four years, resulting in a foreign exchange outflow of approximately US$600–800 million per annum. Under these circumstances, any shift from petrol to diesel is a shift which is not desirable from the viewpoint of society.

Diesel cars in Sri Lanka. A similar distortion in consumer choices in automobiles has been noticed in Sri Lanka. Up to 1979, retail petroleum pricing policy was characterized by subsidies on kerosene and diesel sales, which were partially offset by a high gasoline price but which still resulted in large net losses (Rs 630 million in 1979) to the Ceylon Petroleum Corporation.[37] The ratio of auto diesel to gasoline prices was 0.36 in early 1980, which encouraged consumers to take advantage of this differential by importing diesel cars.[38] While the purchase price of diesel cars is generally somewhat higher than comparable gasoline vehicles, this has been more than compensated by lower fuel costs. As a result, the proportion of diesel vehicles in new car registrations rose from 14 percent in 1978 to 38 percent in 1980. This resulted in resource misallocation because the retail price differential for diesel and gasoline far exceeded their relative opportunity costs; private benefits from switching to diesel cars would exceed the benefits to the economy as a whole. This also had an impact on refinery imbalance problems in Sri Lanka where there was a surplus of gasoline/naphtha on the order of 130,000 tons in 1980 and a deficit (and consequent import) of 42,600 tons of diesel. In an attempt to reverse the trend towards diesel vehicles, the Government of Sri Lanka raised the price of diesel in 1980–81 to 60 percent of gasoline prices and, in November 1981, it revised the licence fees for private diesel automobiles to three times the level for comparable gasoline cars. With the available data, it is not possible to estimate the relative economics of using diesel cars for private consumers. Obviously, a diesel car would still be economical where the utilization rates are high (e.g., intercity private taxi). However, there is a need for a thorough analysis of these options which takes into account the relative prices of import of diesel, export of naphtha, and the fuel efficiencies of different types of automobiles.

Electricity pricing and consumer choices in irrigation pumping

As discussed elsewhere, electricity in rural India is substantially subsidized. The rates charged from the farmers are much lower than the marginal costs of providing electricity to the farmers. Besides, a farmer is not required to pay the costs of connection of the well or tubewell to the nearest source of supply. All these costs are borne by the State Electricity Boards as part of the rural electrification programmes to encourage use of electricity for irrigation. On account of these concessions, the farmers have shown a preference for electric motors to diesel engines. Although the number of electric motors and diesel engines were the same in the early 1960s, the number of electric motors has increased much faster than the stock of diesel engines. In March 1983, there were about 4.5 million electric pumpsets compared with around 3 million diesel pumpsets.

However, the farmers' preference for electric motors is based on cost advantage apart from mechanical problems with diesel engines. Of the 76 nonbeneficiaries surveyed who were willing to install pumpsets, 56 percent opted for electric pumpsets on account of lower capital and operating expenses.[39] The cost advantage of the electric motor (receiving electricity from the regional grid) for pumping water for irrigation[40] is shown in Table 5.2. At market prices (i.e., when viewed from the viewpoint of the farmer), the capital costs of the electric pumpset are almost the same as that of the diesel engine. Since electricity is subsidized (2 U.S. cents/kWh) for the farmer, annual operating costs for the electric pumpset are very low (Table 5.3). Under certain assumptions, use of the electric pumpset is the cheapest option involving an expenditure of US$671 in present value terms.[41] (Table 5.4.). The next best option is use of a diesel pumpset where the present value of costs at market prices is estimated as US$1,555. Under these assumptions, installation of a biogas plant or gasifier to substitute for diesel oil is found uneconomical. The present values of costs are much higher[42]: US$1,766 and US$2,122.

However, the relative ranking of alternative technologies changes when analysis is done using shadow prices in place of market costs. Use of an electric pumpset involves infrastructure costs[43] relating to cost of connection (US$500) and costs of electricity generation and transmission (US$800). Inclusion of these costs shows that the analysis is from the viewpoint of society and not from the viewpoint of the farmer (consumer) who may not have to pay for these costs of connection and generation. Similarly, costs of energy (electricity and diesel) shown in Table 5.3 also reflect shadow prices or real resource costs of using these resources in the economy. For electricity the estimated shadow price (including fuel and operating costs, but excluding capital costs) is 7.48 US cents/kWh[44] compared with a tariff rate of 2 US cents/kWh. Similarly, the shadow price of diesel oil has been calculated by taking the c.i.f. price of imported diesel oil, multiplying by 1.25 to reflect the 25 percent premium on foreign exchange, and adding average local transport costs.

When shadow prices (instead of market prices) are used, the relative ranking of various alternatives changes. The capital costs of using the electric motor when electricity is supplied from a regional grid amount to US$1,700 (Table 5.2). In present value terms (10-year life, 10 percent discount rate), the cost of the grid electricity alternative is much more expensive than the use of diesel engines or dual-fuel engines (using biogas and diesel). Thus, use of a diesel engine is more economic   from the viewpoint of society when all the costs associated with the use of electric motors are included and electricity is valued at its (unsubsidized) shadow price. However, as mentioned earlier, use of the electric motor has been artifically made cheaper by giving subsidies. This has resulted in distortions in consumer choices under which a large number of connections for electric motors have been given, even though power generating capacity to meet these demands has not been added. This has resulted in (1) excess demand for power for agricultural purposes, (2) unreliability of supply along with voltage fluctuations,[45] (3) a tendency on the part of farmers (consumers) to invest in "back-up" systems which result in overcapitalization, (4) disincentives for investing in biogas plants and/or gasifiers which can be used for water pumping, and (5) disincentives for energy conservation measures. The costs of back-up systems have already been discussed in an earlier section. The disincentive effects on renewable energy sources and conservation are discussed below.

Adoption of renewable energy technologies

Although renewable energy technologies such as biogas plants, gasifiers (along with dual-fuel engines), photovoltaic pumping systems, and windmills can be used in place of electric or diesel pumpsets, these technologies have not become popular mainly because of the subsidies on conventional sources, (e.g., electricity, diesel, and kerosene). Use of a family-size biogas plant is not found economic   because the saving in cash for the quantities of kerosene replaced in lighting[46] is not sufficient to cover costs. In the case of irrigation pumping, use of a biogas plant along with a dual-fuel engine (using 70 percent gas, plus 30 percent diesel) is found economical from the viewpoint of society. Its costs (present value terms) are marginally higher than the costs of using a diesel engine (Table 5.4) but are substantially lower than the electricity alternative. However, this advantage disappears when market prices are taken into account. Given the subsidies in capital and operating costs of using an electric motor, the farmer does not find it economic   to install a biogas plant for irrigation pumping. Thus, distortions in energy prices result in nonadoption of technologies based on renewable energy sources. This disincentive has resulted in low diffusion of these technologies and is not considered desirable from the viewpoint of society.

Kerosene price subsidy in Sri Lanka[47]

In Sri Lanka, there is a general subsidy on kerosene whose sale price (US$0.82/gal) is lower than the refinery price (US$1.21/gal). This subsidy has resulted in a revenue loss of approximately US$28 million in 1981 and an encouragement to use kerosene as heating fuel. The main rationale for lower kerosene prices is the government's concern about the welfare of poorer households for whom kerosene is an important lighting fuel. Although this is a valid social objective, the general subsidy on its price may not be the most efficient way of achieving it, since the subsidy is for all users of kerosene and not just for poor households. It has been estimated that a large proportion of kerosene consumption (well over half) is in uses for which the subsidy was never intended — such as industrial heating fuel, in stand-by generators, or as cooking fuel for the better-off households. Besides, the differential in prices of kerosene and other fuels constrains the ability of the Ceylon Petroleum Corporation to raise fuel oil prices. Furthermore, by raising the price of diesel fuels to cost-covering levels, the differential between kerosene and diesel fuels has widened significantly, encouraging the blending of kerosene with these fuels. The issue of reconciling the objectives of equity and efficiency is discussed in a later section.

Prices of LPG and gasoline in Bangladesh

A similar possibility of substitution of LPG for gasoline exists in Bangladesh[48] on account of the price differential between the two products. For example, LPG prices were only 28 percent of the premium gasoline prices on an equivalent energy basis. If this level of financial incentive is continued after LPG availability increases (from natural gas), a significant amount of this fuel could be used to substitute for gasoline in spark-ignition engines, which after small modification (costing about US$300) can burn LPG and gasoline as dual fuels. Even if LPG is priced at its full opportunity cost, it would still be about half of the gasoline retail price at an equivalent energy basis. This preference for LPG against gasoline will further exacerbate the gasoline/naphtha surplus in the country. Besides, given the differential between diesel and gasoline prices, the financial benefit to a private motorist in buying a diesel car is much higher than the economic benefit to the country. This also requires a review of gasoline and diesel prices in Bangladesh.

Disincentives for energy conservation

One of the adverse effects of subsidized energy prices is that the consumers have no incentive to improve the efficiency of utilization of energy inputs. It has been estimated that about 15 percent savings in energy consumption in developing countries could be achieved by a programme of demand management designed to increase the efficiency of energy use.[49] About half of this saving can be achieved through retrofitting and technical improvements in industry, electric power generation, and transport. In India, assessments of potential fuel savings are around 20 to 25 percent in industrial units, automobiles, agricultural pumpsets, etc. A part of these savings could be achieved through improved housekeeping measures, but significant savings would require investments and organizational support to carry out these changes. If energy prices are artificially depressed, the financial profitability of investments in improved utilization is very low on account of low cash savings. This incentive is further reduced if fuel costs form a relatively small proportion of the total cost of production,[50] and if the entrepreneurs are in a position to pass on the high costs to consumers.

In a study[51] of conservation of light diesel oil and electricity used in pumpsets for lift irrigation in Gujarat State in India, it was found that the total investment in rectification (replacement of pipes, foot valves, etc.) of 25 pumpsets in five villages was US$1,283, and the estimated annual saving was US$2,089 for diesel and lube oils. The corresponding costs and savings for 25 electric pumpsets were US$2,310 and US$2,082, respectively. These figures show that energy conservation would be financially profitable and economically viable if estimates of savings are correct. However, the estimated savings are not realized by the farmers, and this leads to uncertainties in financial returns. Hence, lower energy prices through lower cash savings may not provide the same level of benefits which may be available from alternative uses of funds and, as a consequence, may not lead to desired levels of energy conservation.

ENERGY PRICES AND CONSIDERATIONS OF EQUITY AND EFFICIENCY

As discussed earlier, one of the objectives of providing subsidies in fixing energy prices is that these items of basic need should be available to poor people at prices they can afford. This is a laudable objective if it can be ensured that the target groups (rural and/or urban poor) are, in fact, getting the benefits of these subsidies and lower prices. Invariably, this equity objective is not met for a variety of reasons which need to be investigated thoroughly so that the benefits of low prices are available to those for whom they are intended. Besides, it may be necessary to consider measures other than energy prices to provide intended benefits to target consumers. Some of these issues are discussed below, again giving examples from India.

Intended benefits do not reach the poor

The equity objective of the energy pricing is not achieved since the intended benefits do not reach the poor for a number of reasons discussed below:

1. The actual price paid may be higher than the price fixed by the government on account of high distribution margins, ignorance of the consumers, method of purchase and use, and overall shortage of supplies. In India, the kerosene retail price is fixed by the government at Rs 1.80 (US$0.18) per litre to be sold through government-run ration shops. Invariably, adequate quantities are not sold through the ration shops and consumers have to make purchases at higher prices in the black market. In rural areas, the effective price paid by the illiterate villager is much higher since he is purchasing kerosene for lighting in a small lamp every two to three days. During periods of shortages (which are very frequent), the poor consumer has to do without his quota of kerosene and has to depend on much more expensive vegetable oils. Since kerosene is cheaper (US$0.18/litre compared with US$0.32/litre for diesel), there is an incentive for people to mix it with diesel or petrol in trucks, buses, and auto-rickshaws.[52] Under these circumstances, adequate quantities at fixed (low) price are rarely available to the poor consumers, especially in rural areas. Thus, the apparent reason for keeping kerosene prices low is to help these poor consumers, and the issue of kerosene price has become politically very sensitive.[53] However, for policy planners it is necessary to ascertain, through survey, facts regarding the following: (1) the quantities of kerosene actually purchased by the poor in rural and urban areas, and (2) prices *actually* paid by them after taking into account spillage, short measurements, interest component, and value of items paid in kind. By such a survey, the policy planners can educate the politicians and suggest some alternative measures (e.g., supplying free lamps) that directly benefit the poor people instead of subsidies on kerosene which the poor actually do not receive.

The case of subsidized electricity in India is also very similar. Since the costs of wiring a house for electricity connection (US$20–30) are very high for the poor people (whose cash incomes are very low), they are not in a position to take advantage of domestic connection. For many of them, even a monthly charge of Rs 5 (US$0.5) would be quite high, considering other basic necessities. Besides, many poor families have thatched houses where wiring cannot be done or would expose the house to risks of fire. As a result, only about 18 percent of the households in rural areas of electrified villages have household connections. These are, invariably, better-off sections of the rural society, and these people are, in fact, getting the benefits which are being justified in the name of the poor. In this context it is necessary to redefine the scope of the rural electrification project so as to include housewiring as a part of the rural electrification project and charge a nominal amount per month. In fact, a scheme on these lines has been successfully implemented in Tamilnadu in south India. Each of the

100,000 households included in the scheme was electrified at the cost of the Tamilnadu Electricity Board and a connection was given for one bulb (40 watts). Power consumption charges are being collected at Rs 2 per household per month. Although there have been some unauthorized extensions of power supply under the scheme, it has been a boon to poor households who cannot afford to spend the initial amount for wiring. Such schemes may be formulated and field tested to find ways and means that the large masses of poor people can benefit from infrastructure which has already been laid down, and from social costs, which have already been incurred. Besides benefiting the poor, the scheme would also reduce consumption of imported kerosene. Another way of providing benefits from electricity would be to make community services (TV, radio, study room, recreation room) available to the common people, which would provide them with greater social benefits than subsidized electricity which they do not use.

Benefits, in fact, accrue to the rich

Although energy prices are subsidized with an objective of providing benefits to the poor people, the socio-economic conditions prevailing in the country may be such that benefits, in fact, accrue mainly to the rich. This is generally true of electricity subsidies in rural areas where the beneficiaries are rich farmers and artisans who could afford to pay higher prices. Some evidence regarding the beneficiaries of rural electrification programmes in India was collected by the Committee on Power (1980) which concluded that while the rural electrification programme had brought economic benefits to the rural areas, it suffered from: (a) a bias in favour of the large farmer and inadequate availability of rural electrification benefits to the small and marginal farmers, (b) poor progress of domestic lighting and street lighting programmes. A recent evaluation by the Planning Commission (Government of India, 1983a), shows that the low-income beneficiaries (income less than US$100 per year) constituted less than one percent of the total.[54] Another 11 percent of the total sample beneficiaries of rural electrification programmes had annual incomes ranging between US$101 to US$250. High-income households (above US$1,000 per year) accounted for 28.8 percent of the sample beneficiaries. Data on consumption of electricity in rural northern India[55] for 1975–76 show that, although low-income households constituted 22 percent of the total number, households using electricity were only 1.8 percent. In high-income categories (US$1,200 and above), which constituted less than 2 percent of households, almost 45 percent of the total households consumed electricity. The level of consumption of electricity also increased with income — 2.8 to 5.7 kWh/year for low-income households but as high as 45 to 124 kWh/year for high-income households. These data clearly show that it is the relatively better-off households who are getting the benefits of subsidized electricity in rural India. Data available will clearly show that the benefits of rural electrification in terms of energization of pumpsets has primarily gone to medium and large farmers.

In the case of private taxi operators who benefit from the retrofitting of diesel engines (see earlier section), the benefits of low diesel prices are not passed on to the consumers. The taxi operators fix their charges on the basis of petrol-driven vehicles, and those who install a diesel engine get the benefit from lower excise duties on diesel compared with those on petrol. In fact, high excise duties on petrol are paid by only a few car users who are not in a position to pass on their costs to others[56] or by middle-income urban dwellers who have no alternative but to hire taxis on important occasions (emergencies, intercity travel). High petrol prices have certainly reduced the profitability of investment in city taxis and have worsened the taxi service in urban areas.

Energy prices and macro dimensions of equity. Recently there have been some attempts at estimating the equity and efficiency implications of energy prices in a macro framework. Kadekodi (1984) has formulated a model to derive pricing rules for public intermediate energy services such as coal, crude petroleum, and electricity. Optimizing pricing rules have been derived for these energy sources and estimated using data from Indian plan models. The results of the model show that for the coming years both electricity and crude petroleum should be brought under a higher tax net and that coal should continue to have a price subsidy. Particularly, the noncoking coals should be subsidized substantially, whereas coking coal prices should be subsidized marginally. Kadekodi further suggested that by working out those tax and subsidy implications, it might even be possible to consider a common pool of government revenue-expenditures for all such energy sources and energy pricing with such a balanced tax-subsidy approach would be welfare-improving.

Murty (1984) has developed a framework for estimating the social costs of alternative sources of government revenue. Using this framework, he finds that the welfare gains of price subsidies are highest for fuel and light, followed by food grains, edible oils, sugar, and clothing. The estimates of welfare losses of increased prices of fertilizers, cement, and electricity show that any reform in public sector pricing in terms of increasing the prices of cement and giving price subsidies to fertilizers or electricity is a welfare-improving reform in India. These studies emphasize the importance of considering energy prices, taxes, and subsidies in an overall macro framework.

RECONCILING OBJECTIVES OF EQUITY AND EFFICIENCY

As discussed earlier, energy prices are fixed at low levels in order to provide fuels/electricity to population for meeting their basic needs of lighting, cooking, and space heating. This objective of providing fuels at prices which people can afford is considered very important by the governments, and the issue of raising prices (e.g., kerosene or electricity) has serious political implications. Sometimes, lower prices (e.g., electricity in India) are justified on the grounds of equity (to help the small farmers) as well as efficiency (to encourage use of groundwater for raising agricultural output).

However, low prices neither achieve the objectives of equity nor promote efficient use of scarce resources. The benefits meant for the poor do not reach them either because of middlemen's profits (kerosene) or because they do not have the necessary equipment (internal wiring of houses or electric motors). Nevertheless, subsidized prices result in wrong signals to the consumers, encourage inefficient interfuel substitution, and discourage conservation. Low prices lead to low profits, low investments, and shortages of energy which have serious implications for economic and social development of the country. Hence, it is necessary to consider policy measures which can reconcile the various objectives of energy pricing; namely, meeting the basic needs of poor people, avoiding misallocation of energy inputs, and raising adequate resources for investments. Some ideas regarding these issues include: (1) A comprehensive, integrated price policy for energy inputs and other factors of production (labour, capital) should be formulated; (2) schemes of providing direct subsidies to target groups rather than a general subsidy should be evaluated; (3) public investments in the provision of energy should enable consumers to obtain full benefits; (4) complementary inputs and finances should be provided to enable target consumers to avail themselves of the benefits; (5) subsidies may be given on the cost of equipment rather than on fuels; (6) explicit subsidies should be provided for renewable sources whose adoption is adversely affected by subsidized prices for conventional fuels; (7) taxes should be levied on equipment and consumers which take undue advantage of the lower prices (e.g., diesel automobiles, kerosene generators).

Need for comprehensive, integrated pricing policy

There is a need for an integrated pricing policy which covers all energy sources, as well as labour and capital. For example, the policy should cover all relevant energy sources for various end-uses — for lighting it should consider kerosene, grid electricity, biogas, photovoltaic lighting, and electricity from decentralized sources (e.g., microhydro, large-scale biogas, or windmills); for cooking it should consider fuelwood, charcoal, soft coke, crop residue, animal residue, biogas, kerosene, and LPG as possible sources, and relative prices should be fixed to encourage the desired type of interfuel substitution. The objective of equity cannot be met if all the fuels are not considered simultaneously as has been illustrated by the policy of pricing urban fuels in India.[57] Kerosene, electricity, and LPG, which are available at subsidized prices, are used primarily by higher income groups.[58] The subsidy on soft coke, which was primarily used by low-income people, was rather marginal. Fuelwood, agricultural wastes, etc.

which were purchased by the urban poor have not attracted any subsidy or price controls. The result is that the urban poor have to pay high prices for smoky, low-efficiency fuels, while the advantages of subsidized cleaner fuels are available to the better-off consumers. Such anomalies can be removed only when a comprehensive policy is formulated to cover various end-uses (lighting, cooking, irrigation pumping) and alternative fuels for target groups (urban poor, urban middle class, rural poor, rural middle class, small farmers). There is a need for a comprehensive study of the economic costs of providing these fuels to various target groups for different end-uses. The results can then be used as a reference point for determining the tax and subsidy levels for different fuels, different equipment (e.g.,lanterns, bulbs, pumpsets, stoves, biogas plants) and different target groups (e.g., kerosene coupons for the poorest).

There is a need to keep in mind that overenthusiasm in raising energy prices should not result in another form of distortion relating to interfactor prices. In many developing countries, the existing prices of capital, labour, and foreign exchange may not necessarily reflect their "true opportunity costs" due to market imperfections and/or administrative controls (minimum wages, differential interest rates, import quotas, exchange controls). Given the distortions[59] in prices of non-energy inputs (whether justified or not), fixing energy prices which reflect their "true opportunity costs" may, in fact, provide wrong signals to the consumers and result in substitution of scarce nonenergy inputs (capital or foreign exchange) for energy inputs. These considerations point to the need for a comprehensive analysis of energy prices in an integrated framework as outlined by Munasinghe (1980).

Direct subsidies for target groups

Rather than provide a general subsidy, the governments can consider giving direct subsidies to target groups through special coupons. One such experiment has been under way in Sri Lanka. Apart from providing subsidized kerosene to all, the Government of Sri Lanka simultaneously operates a kerosene stamp scheme under which roughly the poorer half of the population (about 1.5 million families) receive monthly coupons, which can be used to pay for kerosene or basic food products. This scheme was introduced in 1979 when kerosene prices were tripled and the value of the coupons was fixed at that time to enable households to purchase about 6 litres of kerosene per month at no additional cost. Since then kerosene prices have increased and the value of the stamps has remained fixed so they no longer entirely offset the higher cost of kerosene, but the stamp scheme can easily be modified to reflect these developments.

In principle this seems to be an attractive scheme because it removes the general subsidy and makes it available to target groups only. The potential advantages of the scheme[60] are: (1) as kerosene prices would no longer be below their economic cost, the incentives to use this fuel inefficiently would be reduced; (2) this would give the Ceylon Petroleum Corporation more freedom to alter the prices of competing fuels, such as diesel and fuel oil, whose prices have had to be held down for fear of diverting demand to subsidized kerosene; (3) by replacing a subsidy in kind (cheap kerosene) with an effective cash transfer, the welfare of poorer households would be increased to the extent that they would choose to spend this higher income on other goods upon which they placed a higher value; (4) the refinery balance problem would be alleviated to the extent that kerosene consumption was reduced as a result of this policy.

Although the scheme seems very attractive to reconcile the objectives of equity and efficiency, it may be too soon to come to a conclusive judgement on this issue. It may be necessary to do a comprehensive review of the kerosene stamp scheme, including its administrative costs and problems and the impact it is expected to have on interfuel substitution.

Extending the scope of public investments

Although the objective of subsidized electricity was to help the poor people in rural areas, they could not obtain the benefits of rural electrification programmes since they could not afford the initial costs of internal wiring. It may be worthwhile to enlarge the scope of rural electrification programmes to include internal wiring of houses at the government's expense to enable the consumer to use electricity. Though such an inclusion would add to the costs of rural electrification schemes, it would provide benefits of better lighting to the consumers and result in a more equitable distribution of benefits from large investments in rural electrification infrastructure. A scheme on these lines has already been introduced by a state government in South India as discussed in an earlier section. There is a need to evaluate this scheme and draw policy conclusions from its work.

Subsidies on costs of equipment rather than on fuels

The governments can achieve the objectives of equity and efficiency by providing subsidies on the cost of equipment rather than on the price of fuel. For example, subsidies may be provided on electric motors, diesel engines, biogas plants, petrol engines, dual-fuel engines, hurricane lanterns, improved kerosene lamps and stoves, and fluorescent tubes to provide incentives for the selection of appropriate (from the viewpoint of society) and energy-efficient equipment and devices. These subsidies could take the form of general subsidies and be introduced through reduction in excise duties on those items. The subsidies could also be special purpose and directed at a particular target group for which elaborate administrative arrangements would have to be made. For example, it would be more equitable and efficient to provide fuel-efficient lamps or hurricane lanterns to everyone in the rural areas, rather than supply kerosene at subsidized prices at which it might never be available. Similar proposals for improved stoves at subsidized prices can be considered. However, there may be difficulties in implementing some of these proposals because energy-efficient equipment which is subsidized would affect the demand for existing manufacturing units which may be in the small, unorganized sector. Thus, there may be a conflict between providing energy-efficient stoves and pumpsets, and employment (and income) in the small-scale industries.[61]

SUMMARY AND CONCLUSIONS

The foregoing discussion shows that energy prices, if arbitrarily fixed, can have serious implications for investment allocation and consumer choices, resulting in inefficient allocation of resources, wrong type of interfuel substitution, and shortages in supplies. There are ways of reconciling the objectives of equity and efficiency in the context of energy prices, and a few policy alternatives have been suggested. The chapter also emphasizes the need for developing a methodology under which prices of energy and nonenergy inputs are determined in an integrated framework.

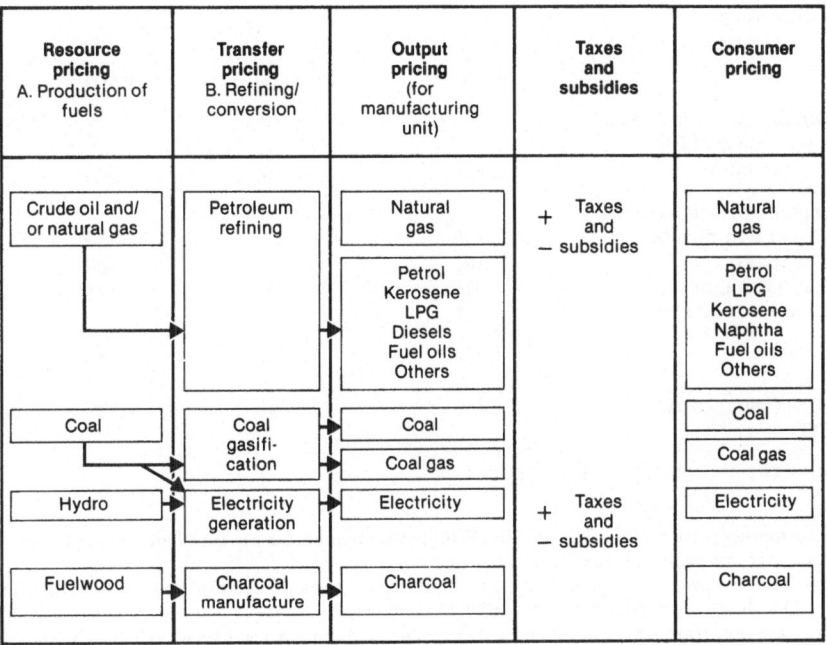

Figure 5.1 Dimensions of energy pricing

Table 5.1 Economics of using diesel engine in a private taxi in India

Costs	At market prices		(1983 US dollars) At shadow prices	
	Ambassador car retrofitted with diesel engine	Ambassador car with petrol engine	Car retrofitted with diesel engine	Car with petrol engine
A. *Capital cost*				
Original price	7,000	7,000	7,000	7,000
Retrofitting of diesel engine	2,500	—	2,500	—
	9,500	7,000	9,500	7,000
B. *Annualized capital costs* (5-year life @ 15% interest rate)	2,834	2,088	2,834	2,088
C. *Annual operating costs*				
Repair and maintenance	350	200	350	200
Fuel costs	1,160	3,100	1,000	1,450
Lubricating oils	116	310	145	387
Driver and cleaner	*	*	*	*
Total	1,626	3,610	1,495	2,037
D. *Total annual costs* (BC)	4,460	5,698	4,329	4,125

* These costs are common to both options.

Notes:
1. Diesel market price is US$0.348 per litre. With diesel engine a car can travel an average 15 kms/litre. Shadow price for diesel oil is US$0.30 per litre.
2. Petrol market price is US$0.62 per litre. A petrol car gives an average 10 kms/litre on intercity routes. The shadow price of petrol is US$0.29 per litre.
3. A private taxi travels an average of 1,000 km/week, giving a total travel of 50,000 km/year.
4. Every two to three years a diesel engine requires overhaul, costing an additional US$300.
5. Cost of lubricating oils is taken as 10 percent of fuel cost.
6. The capital recovery factor (CRP) for 5-year life at 15 percent discount rate is 0.298315.

Table 5.2 Capital costs of prime movers and related equipment for irrigating a one hectare farm in North India

	At shadow prices		(1983 US dollars) At market prices	
	Electric motor (2HP) with grid electricity	Diesel engine or dual-fuel engine (3 HP)	Electric motor (2 HP)	Diesel engine or duel-fuel engine (3 HP)
1. Cost of engine/motor	300	400	300	400
2. Cost of pumphouse	100	0	125	0
3. Cost of connection	500	0	—	0
4. Cost of generation and transmission	800	0	—	0
5. Cost of infrastructure of diesel transport	0	50	0	—
6. Biogas plant (2 m^3/day)	—	400	—	500
7. Gasifier	—	450	—	450
8. Total (12345)	1,700	450	425	400
9. Diesel and biogas (156)		850		900
10. Diesel and producer gas (157)		900		850

Notes:
1. Data are for Ghazipur district in the Gangetic plains of north India. Crop rotation is ricewheat. Water table is taken as 5 M (total head).
2. Cost of biogas plant is US$500, about 60 percent of which is labour costs. Taking the shadow wage rate of 50 percent of the market wage and putting a premium of 25 percent on other materials (steel and cement), the shadow price is US$400. For gasifier, labour costs are one-third of total.
3. Diesel and biogas includes cost of diesel engine (15) plus cost of biogas plant (6).
4. Diesel and producer gas includes cost of diesel engine (1+5) plus cost of gasifier (7).

Table 5.3 Annual operating costs of alternative technologies at shadow prices and market prices

	At shadow prices				At market prices			
	Electricity from grid	Diesel oil	Diesel + biogas	Diesel producer gas	Electricity from grid	Diesel oil	Diesel + biogas	Diesel+ producer gas
1. Energy/fuel cost								
—Electricity	40	—	—	—	10	—	—	—
—Diesel	—	91	44	27	—	106	51	31
—Lube oils	—	8	6	4	—	7	5	3
—Fuelwood	—	—	—	30	—	—	—	30
—Cow dung	—	—	—	—	—	—	—	—
2. Repair and maintenance cost of electric motor/diesel engine	30	45	45	45	30	45	45	45
3. Repair and maintenance cost of biogas plant or gasifier			10	68			10	68
4. Labour charges for operation		30	30	30		30	30	30
5. Total annual operating gross	70	174	135	204	40	188	141	207
6. Present value of operating costs (assuming 10-year life and 10% discount rate)	430	1,069	829	1,253	246	1,155	866	1,272

Conversion ratio: US$1 = Rs 10.00

Notes:
1. Source: See Swiss Development Cooperation Agency (1985).
2. Diesel consumption 0.45 litres/HP-hr for diesel engine: 0.135 litres/HP-hr for dual-fuel engine.
3. Market prices: Electricity 2 US cents/kwh: diesel oil US$0.348/litre. Shadow prices: Electricity 7.48 cents/kWh; diesel US$0.3/litre. The cost of fuelwood is taken at US$40/ton, both at market prices and shadow prices.
4. It is assumed that a dual-fuel engine will use 0.135 litres of diesel per HP-hr along with 1.1 kilogram fuelwood (assuming 3.5 kilograms of wood replaces 1 litre of diesel.
5. Cost of cow dung is not included since the slurry issued from the biogas plant is at least as good a manure as cow dung (if not better). Hence, there is no resource cost in using cow dung as input.

Table 5.4 Sum of present values of capital and operating costs for irrigation alternatives

	At shadow prices		(In 1983 US dollars) At market prices	
	Costs	Rank	Costs	Rank
Electricity from grid	2,130	III	671	I
Diesel oil	1,519	I	1,555	II
Dieselbiogas	1,679	II	1,766	III
Dieselproducer gas	2,153	IV	2,122	IV

NOTES

[1]For example, in India, it would be useful to fix a high price for kerosene/diesel and a very low price (about zero) for fuel oils to encourage refineries to set up secondary processing units (such as hydrocracking units) to convert fuel oils into kerosene/diesel. For details, see Bhatia (1976, 1983).

[2]In the report of the Oil Prices Committee (1976) in India, the need to keep overall prices of petroleum products within manageable limits was cited as one of the arguments for setting the price of domestic crude oil lower than the import price.

[3]For details, see World Bank, (1982b).

[4]Subsequently, there have been price revisions in July 1979, February 1981, and March 1984. The February 1981 price adjustment raised noncoking coal prices by 20 to 30 percent. Even this increase aimed at meeting only the average operating costs of the sector, plus depreciation and interest, and did not provide for any significant return on capital.

[5]Data refer to Coal India Limited (CIL), which covers five operating companies and accounted for 88 percent of coal production in India in 1980–81. See World Bank (1982b).

[6]As shown earlier, this was done by (1) exclusion of return on capital and/or depreciation from average cost of production estimates, and (2) inadequate allowance for increases in cost of inputs between price adjustents that varied from one to four years.

[7]Between 1969 to 1970 and 1977 to 1978, India exported around 0.5 million tons to its neighbours. Due to shortages of coal, exports declined to 0.1 million tons in 1979–81.

[8]See World Bank (1982b).

[9]The calorific value of coal is taken at 5,000 kcal/kg and that of fuel oil as 10,000 kcal/kg. The c.i.f. price of fuel oil in 1981 was US$194 per ton.

[10]See Siddayao (1981a) and World Bank (1982b).

[11]Imports of crude oil and petroleum products at about US$5.5 billion accounted for 65 percent of India's total exports in 1982–83.

[12]See Siddayao (1981a) for a detailed discussion of this argument in the contexts of coal and natural gas pricing in the Philippines and Thailand.

[13]Total consumption of fuel oils increased from around 5.8 million tons in the mid-1970s to around 7.2 million tons in the early 1980s, an increase of around 25 percent or 1.4 million tons per annum. Around half of this increase was being met through imports, involving an annual outflow of US$100 million. This amount could be saved if adequate supplies of coal were made available.

[14]A study carried out by Coal India Limited in 1976 shows that a 20 percent increase in the price of coal will increase the general price level by only 0.5 percent. Even if the impact is higher for certain consumers, the advantages and disadvantages of increasing prices have to be looked at in the context of the total changes in the economy, i.e., including the costs of imported fuel oil.

[15]It is sometimes argued that Indian coal cannot be exported due to low quality and uncertainty of supply. This argument is based on short-term considerations. India has large reserves of high-quality (greater than 6,200 kcal/kg), low-sulphur coal which could be developed for export and has the necessary rail and port facilities. The current shortages of coal in India are the result of wrong pricing policies and investment, which can be corrected to provide surpluses for export.

[16]See Siddayao (1981a).

[17]Conversion factor used here is Rs 10=US$1.

[18]See Government of India (1983b).

[19]See Government of India (1983c).

[20]In 1981–82, 1.969 million tons of kerosene were imported at a cost of Rs 6,065 million, giving a c.i.f. price of Rs 2,400/kilolitre (kl). The market price of kerosene was Rs 1,800/kl (or US$180/kl).

[21]Total subsidy on kerosene, an imported fuel, amounts to more than Rs 2500 million (US$250 million), while that on soft coke is only around Rs 80 million.

[22]At present, the open stack system is used which is inefficient and polluting.

[23]Assuming that an annual consumption of 3.25 million tons would have continued over the six-year period, giving a total demand of 19.5 million tons. As against this estimate, the actual demand during these six years was only 13.9 million tons.

[24]Assuming that (1) soft coke has 6,300 kcal/kg with an appliance efficiency of 22.2 percent, giving 1,400 kcal/kg of useful heat, and (2) kerosene has 10,000 kcal/kg with an appliance efficiency of 50 percent. This gives an equivalent kerosene import of 1.56 million tons.

[25]See Government of India (1980).

[26]For details, see S. Ramesh, (1980).

[27]See Gellerson (1979), p. 175.

[28]See Government of India (1983b).

[29]See Ramesh (1980).

[30]Internal resources include general reserve and depreciation funds, deposits from consumers, employees' provident funds, etc.

[31]See Government of India (1979).

[32]The available pumpset may be used for purposes other than irrigation. However, the main motivation of purchasing and maintaining a diesel system may be for providing timely irrigation when the electric pumpset does not work.

[33]Total number of villages electrified in the population group of above 500 was approximately 150,000. Assuming at least one industrial unit of 10 HP (horse power) each per village, it gives a total capacity of 1.5 million HP. Assuming that about 33 percent of the units keep back-up systems, this gives an estimated capacity of 0.5 million HP costing about Rs 50 crores at Rs 1,000 per HP.

[34]The petrol engine is not sold but is kept as a back-up system when the diesel engine is under repair and overhaul.

[35]The shadow price of petrol has been equated with another light distillate, naphtha. If, at the margin, a ton of petrol is not consumed, it would result in a reduction in naphtha imports by one ton, or an increase in naphtha exports by one ton.

[36]The demand for high-speed diesel oil has been increasing at the rate of 8 to 10 percent per annum.

[37]For details, see World Bank (1982c).

[38]Unlike in India, Sri Lanka car imports were liberalized during this period. If diesel car imports were allowed in India, a large number of private vehicles would be imported to take advantage of the price differential.

[39]See Government of India (1983a), p. 102.

[40]Data are for a representative 1-hectare farm in Ghazipur district of Gangetic plains in north India.

[41]Assuming 10 percent rate of discount and a 10-year life.

[42]The corresponding costs of other renewable energy sources (e.g., solar photovoltaic or solar thermal pumping systems and windmills are even higher. See Swiss Development Co-operation Association (1985).

[43]For details, see Swiss Development Co-operation Agency (1985).

[44]For details, see Swiss Development Co-operation Agency (1985).

[45]The recent evaluation of the rural electrification programme shows that of the 817 beneficiaries, 87 percent reported interruptions in power supply and about 54 percent felt that the damage to agricultural production was severe. About 93 percent of the beneficiaries reported voltage fluctuations, and 38 percent of the electric motors were damaged owing to voltage fluctuations.

[46]In rural areas, fuelwood, crop residues, and animal residues which do not involve any cash expenses are used for cooking. Therefore, the only saving from a biogas plant is in terms of kerosene used for lighting.

[47]See World Bank (1982c).

[48]See World Bank (1982a).

[49]See ESCAP (1982), p. 60.

[50]In 1975–76, the average incidence of energy costs as a percentage of the value of production in industry was only 6.8 percent. Industries where incidence of energy costs was higher than 15 percent accounted for only 9 percent of the total output. See Government of India (1979).

[51]For details, see Patel (1982). Also see Patel and Gupta (1979).

[52]In the late 1960s when the price of kerosene was lower than that of diesel, a substantial quantity of kerosene was diverted and mixed with diesel. See Desai (1979). The retail price of diesel has been higher than that of kerosene since June 1980. The recorded growth rates in consumption of kerosene have been higher during this period: 9.2 percent in 1980–81, 11 percent in 1981–82, and 10.6 percent in 1982–83.

[53]In February 1983, there was an attempt to introduce dual pricing in kerosene: US$0.17/litre from the ration shop and US$0.3/litre in the open market. However, the scheme had to be immediately withdrawn due to political pressure since raising kerosene prices has become an emotional issue in India.

[54]The per capita income for the year of survey (1977–78) was about US$120 or US$600 to US$700 per household. Rural income levels were likely to be lower.

[55]As quoted in Desai (1979).

[56]It is estimated that a large number of cars in India are either owned by the government, government agencies, private sector companies, or their employees. For these consumers, higher taxes have no role to play since they shift costs to the government directly or indirectly (through higher operating expenses and lower direct taxes).

[57]For details, see Sharma (1983).

[58]According to a fuel-use survey, 37 percent of kerosene, 100 percent of electricity (for cooking), and 76 percent of LPG were consumed by high-income groups averaging US$600 or above per year, while soft coke was primarily (56 percent) used by the low-income people.

[59]For a detailed discussion, see Bhatia (1981) and ESCAP (1982).

[60]These are outlined by the World Bank (1982c). According to their estimates, the kerosene price would be raised from Rs 3.9 to Rs 5.4 per litre. The general subsidy would be reduced by Rs 345 million, while the kerosene stamp scheme would be increased by Rs 172 million, giving an additional Rs 173 million to the Ceylon Petroleum Corporation.

[61]A scheme to subsidize improved kerosene stoves was considered in India. However, it was found that this would be resisted by a large number of small manufacturers who would be completely wiped out. Similarly, any subsidy on improved diesel pumpsets would adversely affect the small manufacturers of those engines scattered in various parts of the country.

REFERENCES

Anandalingam, G. (1984). *The Economics of Industrial Energy Conservation in Developing Countries*. Delhi, India: Tata Energy Research Institute.

Bhatia, R. (1976). "A multiprocess multiregional programming model for petroleum industry." *Indian Economic Review* (April).

Bhatia, R. (1980). "Energy alternatives for irrigation pumping." in R. K. Pachauri (ed.), *International Energy Studies*. New Delhi: Wiley Eastern.

Bhatia, R. (1981). Energy Pricing in Developing Countries: An Integrated Framework. (Draft) Cambridge, Mass: Harvard University (February).

Bhatia, R. (1983). *Planning for the Petroleum and Fertilizer Industries: A Programming Model for India*. Oxford, UK: Oxford University Press.

Desai, A. V. (1979). *Impact of Higher Oil Prices on India*. International Labour Organisation (September).

Gellerson, M. (1979). "Marginal cost-based electricity tariffs: Theory and case study of India." *Indian Economic Review* (October).

Government of India (1976). *Report of the Oil Prices Committee*.

Government of India (1979). *Report of the Working Group on Energy Policy.*
Government of India (1980). *Report of the Committee on Power.* New Delhi.
Government of India (1983a). *Evaluation of the Rural Electrification Programme.* Vol. II.
Government of India (1983b). *Mid-Term Appraisal of the Sixth Plan.*
Government of India (1983c). *Report on Coal, Bureau of Costs and Prices.*
Kadekodi, G. (1984). Pricing of energy inputs: Cases of coal, petroleum and electricity. Institute of
    Economic Growth. Mimeo.
Koomsap, P. (1981). *Petroleum Products Pricing and Its Impacts, View from an Oil-Importing Coun-
    try.* Conference of the Federation of ASEAN Economic Associations (November).
Munasinghe, M. (1980). "An integrated framework for energy pricing in developing countries." *Energy
    Journal* (July).
Munasinghe, M. (1984). Energy Supply and Demand Management. Draft.
Munasinghe, M. and J. Warford (1982). *Electricity Pricing, Theory and Case Studies.* Baltimore:
    The Johns Hopkins University Press.
Murty, M. N. (1984). *Distributional Equity, Government Revenue, Prices and Tax Policy.* Institute
    of Economic Growth (March). (Mimeo.)
Patel, S. M. (1982). *Study on Conservation of Light Diesel Oil and Electricity Used in Pumpsets
    for Lift Irrigation in Gujarat,* Ahmedabad, India.
Patel, S. M. and R. K. Gupta (1979). *Study on Conservation of Light Diesel Oil Used in Pumpsets
    for Lift Irrigation in Gujarat.* Ahmedabad, India.
Ramesh, S. (1980). Pricing Policy for Power (December). (Draft).
Schramm, G. (1979). The Economics of Energy Pricing (January). (Draft).
Sharma, S. (1983). *Domestic Energy Consumption in India.* IES Training Programme.
Siddayao, C. M. (1981a). *Fossil Fuel Pricing Policies in the Asia Pacific Region: A Preliminary Assess-
    ment of Some Allocative Implications.* Expanded version of report prepared for the Asian Develop-
    ment Bank 1980 Regional Energy Survey issued in the Resource Systems Working Paper Series
    as WP-81-3. Honolulu: The East-West Center.
Siddayao, C.M. (1981b). *Pricing of Fossil Fuels in Asia: Allocative Implications.* Paper presented
    at the International Atlantic Economic Conference, London, 1981, and issued in the Resource
    Systems Working Paper Series as WP-81-11. Honolulu: The East-West Center.
Swiss Development Corporation Agency, New Delhi (1985). Energy Alternatives for Lift Irrigation.
    Mimeo.
United Nations. (1982). *Proceedings of the Committee on Natural Resources, Eighth Session.* Energy
    Resources Development Series No. 25. Bangkok: Economic and Social Commission for Asia and
    the Pacific.
World Bank (1981). *Energy Pricing in Developing Countries: A Review of Literature.* Energy Depart-
    ment. Paper No. 1 (September).
World Bank (1982a). *Bangladesh: Issues and Options in the Energy Sector* (October).
World Bank (1982b). *India: Coal Sector Report* (September).
World Bank (1982c). *Sri Lanka: Issues and Options in the Energy Sector* (April).
World Bank (1983). *Nepal: Issues and Options in the Energy Sector* (August).

Chapter 6

SHADOW PRICING INDIGENOUS ENERGY: ITS COMPLEXITY AND IMPLICATIONS

Corazon Morales Siddayao

INTRODUCTION

Developing country responses to energy price changes have indicated that price, as a signal of the value of this commodity, has been as marked in influencing choices in the producing sector as in the consuming sector. The development of coal, oil, natural gas, and geothermal resources has become more viable as a result of the dramatic rise in oil prices in the 1970s. In the Asian region, the higher-cost petroleum resource accumulations (relative to those in the Middle East or North America) became economically attractive to foreign investors, although the degree of investor response has varied according to specific country contractual terms.[1] Still, two divergent pricing policy trends have emerged in response to developments in the international energy market. While underpricing of consumer energy products was generally the rule rather than the exception in Asian developing countries in the 1970s, the opposite has been emerging as an approach to pricing indigenous energy resources at the supply point, especially in the net-oil importers (see Table 6.1). The approaches, in place or suggested, may be summarized as follows: (1) At one end are cases where governments are concerned about providing producers the opportunity to reap "excessively high profits" if prices are allowed to rise to import parity levels. Resources tend to be priced below their true economic costs. (2) At the other end are advisers recommending that governments in resource-endowed countries use international oil prices as the benchmark for pricing indigenous coal and natural gas (citing the replacement cost concept).[2] The argument on which this recommendation is based has usually been the avoidance of sharp adjustments to higher traded oil prices when domestic resources run out. Furthermore, underlying the argument for using international prices as a benchmark is the assumption that the domestic resource is a tradable good, and that therefore the reference cost to serve as the "shadow marginal cost" for the indigenous resource is the c.i.f. price of imported oil, also referred to as the "border price."

The preceding chapters have provided the setting for the appropriate place of energy policy and, more specifically, of energy pricing policy in the overall economic policy framework. This paper will focus on the issue of shadow pricing indigenous resources in developing countries. Because energy pricing appears to be an area where government intervention has become dominant, this paper questions the basis for that intervention and the approach to determining the price to any buyer of indigenous energy resources.

It will be useful to start with a definition of the term "shadow price," as it is used in this chapter. It is defined as that value assigned to a commodity or factor of production that contributes to a change in the country's socio-economic objectives through a marginal change in the availability of the shadow-priced commodity or factor. Hence, in the words of Squire and van der Tak (1975):

> . . . the process of shadow-pricing presupposes, first, a well-defined social welfare function, expressed as a mathematical statement of the country's objectives, so that the marginal changes can be evaluated; and, second, a precise understanding of the constraints and policies that determine the country's development, both now and in the future, and hence the existing or projected circumstances in which the marginal changes will occur. (p. 49)

An important point also needs to be added about the definition used. The use of shadow prices presupposes the existence of distortions; shadow prices are not equilibrium prices that would prevail in a distortion-free economy.

The two succeeding sections present (1) the basic premises of shadow pricing and the theory of resources, with special attention to energy; and (2) the issues associated with shadow pricing domestic fossil fuel resources. The concluding section summarizes the fundamental issues raised.

OPPORTUNITY COST, SHADOW PRICES, AND THE THEORY OF RESOURCES

No one will argue against pricing energy products to reflect the social opportunity costs of their consumption and production. The method of calculating those opportunity costs, however, is an issue in itself. What is the alternative cost of consuming a certain commodity? If natural gas, for example, is discovered and produced, what are the related costs of producing and consuming this fuel? How much would this fuel command if employed in alternative uses? Is such alternative use viable? If this fuel is not used, what would be the cost of the alternative fuel? Over the short term? Over the long term? The usual approach to shadow pricing is to start by determining the supply-related (or technical) costs. To this must be added the costs of externalities that are not captured by direct costing. That is, to the supply curve reflecting direct economic costs must be added, where possible and desirable, the additional social costs of producing the commodity.

Indigenous energy resource development has basically been encouraged within the framework of conserving foreign exchange by reducing oil imports. Unfortunately, the dramatic oil price increases in the 1970s appear to have revived what one may refer to as an energy theory of value. As a result, policy planning in developing countries has tended to overemphasize the role of the energy input in production and consumption in dealing with the energy problem, rather than to view the problem in the context of overall economic efficiency. This practice persists, despite the voluminous amount of both published and unpublished work attempting to understand the impacts of changing energy prices on the whole economy.[3]

In the process, a surprising twist has developed — a dichotomy in conceptual approaches to pricing energy resources. The emphasis on shadow pricing indigenous coal and natural gas production (to reflect the replacement cost of these fuels) can be interpreted to mean one of two things: (1) these resources are placed in a category identical to public investment projects, which they usually are not, or (2) an assumption is implied that "market failure" prevents freely set prices from reflecting the true social and economic costs of the resource.[4] Such emphasis on shadow pricing obscures the role of the theory of resources which has been the subject of substantial analyses in addressing the problems of mining nonenergy depletable resources.[5] This theory, with some qualitative modifications, may easily be applied to fossil fuel resources, the primary set of energy resources under focus in the foreseeable future. It would be useful to summarize the relevant features of both notions.

Shadow pricing

Shadow pricing is employed in public investment decisions when market prices are assessed to be distorted. For example, production factor (labour or capital) market prices may not bear a close relation to their opportunity costs because (1) unequal rates of return to capital exist that are not justified by risk differentials, or (2) institutional or environmental factors prevent factor prices from reflecting their opportunity costs. When public investment decisions have to be made under such circumstances or when no bases exist for market pricing inputs and outputs in projects, shadow pricing is the appropriate recourse.

The concept of domestic resource cost is a related notion. The concept relates to measuring the real opportunity cost of producing (or saving) a net marginal unit of foreign exchange in terms of total domestic resources. It can be compared to using some measure of the economy's "real" or "accounting" exchange rate to serve as an investment criterion. It is closely related to the notion of comparative advantage as a basis for international trade.[6] It is therefore a useful notion where the issue of import savings is an important consideration — as in the case of indigenous energy resource development.

A broader definition of shadow pricing is that introduced earlier. As Squire and van der Tak point out, shadow prices will depend on both (1) the fundamental objectives of the country, and (2) the socio-economic environment in which the marginal changes occur. Any change in such objectives will require a change in the estimated shadow price.[7] (These points will be recalled later in the discussion.)

Since indigenous energy resource development is basically viewed within the framework of foreign exchange savings as an objective, the usual assumption is that the shadow price of those resources has to be determined rather than freely set by the market. The issue of whether the resource is tradable or not then arises.

The usual starting point for estimating the marginal opportunity cost of tradable goods, such as petroleum, is to use the international or border price (i.e., the c.i.f. price of imports, or the f.o.b. price of exports, with adjustments for internal transport and handling costs), as other authors in this volume have noted (e.g., Schramm in Chapter 4). Coal and natural gas may be treated similarly, depending on whether or not they are tradables. In cases where the border price is believed to vary significantly with the amount bought or sold, the marginal import cost or marginal export revenue is used. Adjustments are made for internal handling and transport costs. Border prices are used not because to do so is a more rational approach; the essential point is that they represent a set of opportunities open to a country and the actual terms on which it can trade.[8] Goods whose costs fall between the bounds of export and import prices are called "home" goods or nontraded goods. For some home goods, a small change in international price or domestic cost may result in exports or imports. For others there may be no possibility of international trade.

The c.i.f. import price is used to measure the value to the economy of any output from a public investment project that substitutes for imports, since it measures the direct foreign exchange cost prevailing at the time the import is replaced. Any input or output whose value to the economy cannot be measured in terms of its border price should be assessed in relation to its home market price. A disparity may exist between the marginal value or demand price of nontraded goods and their marginal cost or supply price, as a result of market imperfections or differential taxation, direct or indirect. Under such circumstances, several criteria have been suggested by Squire and van der Tak:[9]

(1) if the use of an input in an economic activity reduces the supply of that input to other users, its shadow price should be based on the demand price;

(2) if an input is supplied from new production, the shadow price should be based on the supply price;

(3) if the input is supplied from both sources, the weighted shadow price is used, such weights determined by the elasticities of supply and demand;

(4) where indirect taxes (subsidies) compensate for externalities, and exact correspondence exists between such taxes (subsidies) and the costs (benefits), the shadow price should include such increments (deductions), and vice versa.

The nontraded price of a nonrenewable resource, such as coal or natural gas, will normally reflect both the marginal cost of production and some economic rent or "user cost".[10] In either case, the marginal opportunity cost reflects the shadow-priced economic value of the alternative output foregone because of the increased domestic consumption of that particular energy resource. For energy resources, such as fossil fuels, the shadow price must include rents that may be earned. The market measure of such rent is described below. The shadow value may have to be adjusted to reflect distortions in the capital and product markets brought about by market imperfections or fiscal measures.[11] It has also been suggested that such shadow value be adjusted to reflect the impact on savings and income distribution.[12]

Resource theory and energy resources

Before discussing whether or not the suggested approaches to shadow pricing an energy resource are appropriate, it would also be useful to summarize the relevant features of resource theory. Hotelling's r-percent growth rule has come to be viewed as the "fundamental principle" of exhaustible resource economics.[13] The rule, however, depends upon a number of very stringent simplifying assumptions. Most of the recent theoretical work has been aimed at addressing observed resource price behaviour in the light of changes in markets and technology.

In a free market, the size of the resource base determines the supply price elasticity of the resource and the degree of the divergence of the long-run optimal price from the short run. Where the stock is so large as to approximate infinity — i.e., where marginal use does not affect the import or export levels, and therefore does not affect price in the short run — the long-run average cost curve (and price) would approximate the marginal opportunity cost of producing that resource. This curve may be upward sloping where diminishing productivity of extraction inputs exists.

Where the resource base is small, or output capacity is limited, the demand side is more active in determining the rate at which price will change. A change in demand at the margin could cause exports to increase (if the change occurs externally), or import substitution to take place (if the change occurs domestically). If an increase in internal demand occurs, restriction of the resource to domestic consumption could cause marginal use of the resource or output to reduce export earnings from this commodity, or to increase the import bill if import substitution is not possible. In the small resource base case, marginal opportunity cost rises sharply in the short run and then rises over the long run with the international price.

A third case is possible. Not only could the resource base be small, but certain characteristics of the commodity could preclude its trade. For example, domestic coal deposit(s) could satisfy local requirements. Furthermore, if the quality of the coal is relatively poor, this resource would not realistically become a tradable commodity, given the high transport costs associated with coal trade. In such a case, the net price would still rise, although less sharply than in the second case because of the nature of its market. Because a small base implies eventual exhaustion of the deposit(s) within a shorter time frame than that for a large base, the slope of the price path would, at some point, diverge from that of the marginal opportunity cost of producing that resource, as price changes at a rate faster than the real interest rate. This path would thus eventually reflect a scarcity factor.

As the preceding discussion shows, the common assumption that the net price of an exhaustible resource will rise along with the rate of interest[14] will not hold under certain resource base conditions. This assumption will further not hold (1) if technological progress in extraction outpaces the rise in the rate of interest, (2) if the nature of the discoveries results in different costs of production, (3) if substitutes are in sight over the long run, and (4) under noncompetitive conditions.[15] The prospect that substitutes will be developed causes price to rise at a rate at least as fast as the interest rate, and at best as fast as the interest rate plus a factor representing the conditional probability of substitution. If the stock is small, uncertainty about the future can result in depletion at a rate slower or faster than those under conditions of certainty, depending on the circumstances. (In the case of uncertainty over the future institutional framework in a foreign country, for example, a producer may discount the future heavily and deplete a reservoir at a rate faster than that which it would pursue if it perceived that framework to be stable).[16] In a monopoly, the extraction rate is biased downward. (These two last situations will be recalled in a later discussion of the situation in Asian countries.)

Resource production is distinguished from manufacturing by the presence of resource rents and by the generally higher risks associated with the industry itself. Conceptually, the resource rent accruing to a given deposit is determined by the difference between the cost of production (including the cost of capital) for a given deposit and that for a marginal deposit. It may also be defined as the profits remaining after deducting a producer's income that corresponds to the minimum return necessary to attract investment in new projects. The resource rent element is principally related to the quality of the resource, geological and engineering considerations, location, etc.[17] Resource rents have a time horizon; changes in demand resulting from the appearance of substitutes or by discoveries of new deposits affect such rents. The "user cost" element of economic rent was also touched upon earlier.

The price of a resource determines the amount of rent available to a producer, given cost. This price also serves as an allocative device to signal the value of the resource to society. If a government wants to capture any part or all of the rent arising from the production of the resource, it can do so with an effective fiscal framework.[18]

SHOULD FOSSIL FUEL RESOURCES BE SHADOW PRICED AND PEGGED?

This paper began by citing the increasingly common recommendation to developing country planners to "peg" domestic energy resource prices to international oil prices (i.e., to use international prices as shadow values). It will, therefore, be useful at this point to recall the issues relevant to the arguments for or against such an approach.

Related issues reviewed

The supply base of the commodity determines the price elasticity of the fuel and the degree of the divergence of the long-run optimal price from the short run. One of the problems to be addressed in shadow pricing is the question of capital indivisibilities or "lumpiness." This arises in the transportation and distribution of natural gas, in petroleum refining, and in coal mining. As capacity limits are reached in a growing market, the short-run situation will be one of rising marginal opportunity costs.

When one starts talking about the long term, however, one enters the dynamic sphere and the notion of the discount rate. The marginal opportunity cost may then be expressed in terms of the marginal social cost and the social discount rate. That is:

$$MOC_o = MSC_o + be^{-it}$$

where

MOC_o = marginal opportunity cost at time o

MSC_o = marginal social cost of an activity at time o

b   = benefits foregone in the future as a result of consumption at time o

e   = the natural exponential base

i   = the social discount rate

MSC_o is, of course, sensitive to demand/supply conditions. MSC_o may, in fact, be expressed in terms of the dependence of an economy on the international energy market. That is,

$$MSC_o = P(1 + 1/s)$$

where P=the international market price of oil and s=the supply price elasticity. If supply is infinitely price elastic, the social cost of a barrel of imported oil equals the market price as the second component, $1/s$, approaches zero. As the elasticity of supply decreases, and, therefore, as the second component increases, social cost exceeds market price. This second component may be determined in a net oil importer by the size of its indigenous energy resources, physical access to them, and financial access to internationally traded energy resources. This second element thus measures the social premium that might be placed on developing indigenous resources. MOC_o could, of course, determine the shadow price for the indigenous resource, if this is required. Pegging indigenous energy resource prices is discussed in more detail in a separate section below.

Resource pricing methods can be arbitrary; the price may reflect various goals. The Organization of the Petroleum Exporting Countries (OPEC) argues that the price of its oil should be set at the cost of producing the alternatives to oil, rather than at the direct cost of production. Actually, economic rent and the capture of such rent to fund development goals are at issue. Where the resource is produced domestically and is owned by the government, pricing of the resource domestically marketed by the government can be at any level determined by the government, which can capture the amount of rent it wants. (Furthermore, if, as in the case of oil, the resource faces a relatively inelastic demand in the world market, the government's ability to control allocation of rent to itself is augmented.) The revenues resulting from the production of such resources are usually earmarked to finance general economic development programmes. In such combined operations, what is the main determinant of the economic cost of the resources? Is it the foreign exchange that the resource would earn for the exporting country (i.e., the price the international market will bear)? Or does it include the long-term socio-economic cost of oil importers if the present generation is denied consumption of the resource domestically because the price is too high for their levels of income?

There are also several problems related to shadow pricing, especially of petroleum products, but even of fossil fuels together. One set of problems is the long-run effects on the price of one product or fuel of shifting demands. The long-run effects, of course, are highly dependent on the elasticities of substitution and of price. A high substitution elasticity (a technical factor) implies a high price elasticity (an economic relationship). If both elasticities are high, then over the long term the interplay of demand, price, and supply could result in a shadow price that is totally different from the original one calculated, since one cannot determine *a priori* the final combination resulting from these interactions.

Related to this problem is that of joint costs in production. This situation arises in the upstream stages of petroleum and natural gas exploration and production; it also arises in costing different petroleum refined outputs. The issue of joint costs in the exploration and production of natural gas and petroleum is an old one in the field of price regulation;[19] allocation of costs is highly dependent on the specific demand situation. As for refined products, certain items have international price benchmarks (e.g., kerosene, gasoline, and diesel). The heavier outputs, like residual fuel oil, may be treated as nontradables and, therefore, may be priced without weighting their impacts on export and import levels and the resulting foreign exchange impacts.

Which notion should apply?

Still, the basic issues that must be resolved or satisfied in determining whether government intervention through shadow pricing or conventional resource theory should apply in adopting a pricing policy for indigenous fossil fuel resources are: (1) whether or not the resource development project is a public project, or (2) alternatively, that the "market has failed." (Whether this failure has been caused by government policies in the energy sector or others is another matter.)

Shadow pricing a resource primarily implies that the private sector is not involved in its development; if it is, shadow pricing implies that a determination has been made that the market is not working properly and that, therefore, the government must set prices to improve allocative efficiency. In dealing with this issue, an underlying consideration is whether or not the country desires development of the resource. A second consideration is whether or not the country desires to leave such development in the hands of private investors (domestic or foreign), or if it will develop the resource itself.

Where the government of a country undertakes to develop and produce its domestic energy resource, it may or may not take a public investment approach to pricing. Where it does, its shadow price should reflect the true economic costs of supplying that resource. At the minimum it should reflect direct supply costs, including the equivalent of user costs and a return to the investor that would allow it to reinvest its earnings in other projects. A government may require a public corporation to conduct its operations by the standards set by private firms; i.e., to require this agency to yield returns equal to or higher than the real rate of interest (as is done in Singapore's manufacturing sector). In either case, the shadow price should follow certain criteria (which will be taken up later). At this point, it might be useful to recall the conclusions suggested by the existence of noncompetitive conditions on the rate of extraction, and to be reminded that government production implies that, like a monopolist, it has greater discretion in controlling the rate of extraction than is available to a competitive private producer.

Where resource development is undertaken with private funds, the theory of resources applies. The initial question that must be raised is whether or not the market price reflects the true opportunity costs of supply. If not, the next question to ask is whether or not government policy has been responsible for such divergence, directly or indirectly. The third question is, how can the situation be improved, if necessary.

If the private sector is the producing agent and development is desired, profitability to the investor is an important issue. An adequate return is one that assures (1) positive incentives to invest in other deposits, and (2) adequate financial resources for such reinvestment.

If the market price falls below the true opportunity cost to the investor, there is no assurance that the resource base will be adequately explored and produced. If the reason for this is a price ceiling imposed by government, reevaluation of its policy is called for. If the market price allows an adequate return to the investor plus a rent component, a review of taxation policies is in order if the government wants to capture part of that rent. Taxation policies may reflect inadequate direct taxation of the industry or fiscal policies that result in effective interest rates falling below real rates. In any case, the burden of proof of market failure is on the government, and the need for the adoption of replacement cost as a shadow price will have to be shown.

At the same time, one might ask if the domestic energy resource is the appropriate factor to shadow price. If the scarce resource is the country's foreign exchange, then one must address the issue of how an additional dollar of foreign exchange can be saved by producing the indigenous resource and what the real resource values are of the forgone uses if it is not produced. For, in principle, there are three ways to increase one's foreign exchange reserves: (1) by reducing imports through the curtailment of demand for certain goods, (2) by domestically producing goods or natural resources (in this case, energy resources) that normally create a demand for foreign exchange, and (3) by increasing exports.

Furthermore, producing the energy resource domestically implies the application of some capital and labour resources in its production. Some investment is thus displaced. Hence, the rate of return for this particular project should be higher than the rate of return which the capital resources would have in their next best alternative. In essence, the shadow price of an energy resource, if one insists on taking this approach, must recognize all the elements that are involved and therefore must reflect these elements. As this implies, a significant amount of information is required to arrive at the "correct" shadow price. It is not clear that governments are equipped with this amount of information.

Why peg prices to international oil?

Where indigenous resources exist and have the potential for development, it is clear that pricing should be based on the opportunity cost of not producing the fuel or, alternatively, of using the fuel domestically. The opportunity cost issue is, however, confused quite often with the replacement cost issue, so that the only criterion used for determining the price of the indigenous resource is the import cost of the alternative fuel. The first is a broader concept than the second. To equate them is erroneous and has grave economic consequences.

The basic considerations in identifying the relevant opportunity costs are the actual possibility of export and the accompanying foreign exchange gains from such exports, in addition to the foreign exchange cost of continued importation of the alternative petroleum product. A failure to produce the indigenous resource at the higher shadow price (the cost of oil imports) because the resource becomes non-competitive relative to oil imports implies a loss of foreign exchange.

The cost of importing oil is, furthermore, not limited to the actual foreign exchange losses. There are long-term costs when such loss constrains expenditure on development-related projects. The long-term costs either of reducing demand for other development-oriented goods or of external borrowing are important considerations in weighing the implications of the price chosen or the measure of the opportunity costs. The latter seems extremely essential for many countries, both in South and South-East Asia.

At the same time, if export is really not an economically viable possibility for the indigenous resource — e.g., coal in the Philippines and Thailand, or even natural gas in Thailand — pricing the resource at the international level, or at the level of the replacement fuel (in this case, imported oil), would be keeping the price artificially high.

Furthermore, each commodity has its own market, its own characteristics, and its limitations — and therefore its own price. To align natural gas or coal prices arbitrarily on a Btu basis to import oil prices is to ignore these differences. Even when premiums are allowed for quality differentials, such premiums — if set by an administrative authority — may not necessarily reflect the true economic choice coefficients that the choices of buyers and sellers imply.

The principal disadvantage of pegging prices of indigenous resources to oil import prices is the distortion such policy creates. If price is seen as a coefficient of economic choice, then the price a buyer pays for a commodity is an indicator of that buyer's preference for that commodity. Thus, an artificially high price distorts the allocation of resources, just as an artificially low price would. Standard economic theory suggests that when price regulation alters the production of some good X, other goods related to good X through production technology or market demand will also be affected. The initial policy measure may, therefore, distort economic signals to owners of other resources and serve to alter their production decisions. Raising the price above the economic costs based on the direct supply costs and related social costs also imposes a "tax" on the user. This "tax" could alter the ordering of all related possible expenditures from that which would have existed if the price followed a more natural course.

Evidence exists to the effect that in cases where wage rates are not permitted to reflect differences in relative efficiencies, some workers become unemployable. Similar observations may be made of the use of an important intermediate good such as energy, especially where substitution of energy sources is possible. Furthermore, one could argue that pegging an indigenous energy resource at a higher price, which would then be taxed to capture producer rents, attenuates the rights of private producers to sell freely at a lower market rate that covers all their costs.

Even without taking into consideration the possible increased incentives for investment in a sector with overpriced energy resources, the investment needs in the energy sector of Asia to the year 1990 are already enormous (see Table 6.2). Raising the price of the indigenous resource above its supply cost places the consumer's cost at a level that could further encourage premature development of alternatives. (Figure 6.1 shows comparative costs for such alternatives.) While there may be benefits to be gained from these shifts, the premium from such early shifts does not have to be borne by the developing countries — and certainly not by the Asian net-oil-importers — where scarce financial resources could very well yield higher gains when spent on other socio-economic activities.

If it can be determined that pricing the indigenous resource at market levels is inappropriate because the Pareto conditions have been violated by distortions in the market, the proposal to peg the resource to imported fuel prices would have to show that new equilibrium conditions are indeed promoted by this change. These new conditions might be achieved under certain constraints, but could result in improvement in overall efficiencies under such "second-best" conditions.[20] The more closely the good is related to other goods from the viewpoint of consumers or producers, the less desirable are piecemeal policy measures applied to one good. Since energy is of pervasive importance, its impacts on other aspects of consumption and production need to be taken into account. The mutual interdependence that exists between energy markets and others suggests that a change in the prices in the energy markets will inevitably lead to a change in others. As Warr (1980) notes, "When shadow pricing is . . . applied widely throughout a large . . . sector, . . . its informational problems are compounded. The data necessary for the estimation of the optimal shadow prices are not (locally) observable and the welfare gains potentially obtainable from the use of the correct shadow prices can be eroded by quite small errors in the shadow prices estimated."

One of the economic penalties to a society of an overpriced nontradable domestic resource is the inflationary factor that is unnecessarily imposed on the economy. Artificially raising the price of the indigenous resource has the effect of contributing to the inflation index, and unnecessarily so. In countries where inflation is already difficult to control, an artificially high price for the energy resource compounds that problem.[21] (Table 6.3 shows annual growth rates of inflation in selected Asian developing countries; it also compares the acceleration in the 1970s with that in the 1960s.) So long as producers are satisfied with their market price and believe this allows them to recover their capital investment plus a satisfactory return, that price — which may be below, equal to, or even above the international level, depending on demand/supply conditions — should be sufficient. The important criterion is that such price covers the marginal economic and social costs of producing that resource. There is no guarantee that a market-determined price will be below international levels, but a community should be allowed to take advantage of the lower price differentials its indigenous endowments may provide, when they exist, and not prematurely be burdened with a higher shadow price set by government. As the indigenous resource is depleted and as the proportion of imports rise, the average domestic price should increase as a normal course, barring political intervention or unforeseen techno-economic developments that result in a different price path.

The arguments may be illustrated by referring to Figure 1.3 in Chapter 1 of this book. In effect, what has been suggested to countries by external advisers is that the price path should be AD rather than AJE or at worst AFE. If a domestic energy resource is produced and used, the average of the price for this resource and the imported oil would be somewhere along path IJ. The slopes of IJ and AFE will depend on the domestic resource base and, therefore, how fast the domestic price rises as the stock is depleted. This path could, of course, be disturbed by unexpected discoveries, and a less slowly rising or uneven price path could develop which would be located below AC.

Although this paper focuses on net-oil-importers, the argument against keeping energy prices artificially high could be used just as well in a country with abundant, exportable energy resources, if inflation is already a problem. The usual argument for using the export price as the benchmark, and as a measure of the opportunity cost of consumption is a weak one in the Indonesian case. The trade-offs of a nation include weighing the benefits to be gained from the additional foreign exchange that could be generated from reduced domestic consumption as a result of the high prices against the social costs of magnified inflation as a result of artificially high energy prices. Concern over future supplies is not a good reason for keeping domestic energy prices high at the present. Growing scarcity as supplies are diminished will raise prices eventually but gradually, if prices are allowed to move naturally. In the mean time, alternatives could be entering the market. Again, the basic concept to be kept in mind in this case is the social discount rate of consumption. In other words, what is the value to society of consuming a certain level of petroleum resources at present versus the value of postponing its consumption to some future date?

The argument for pegging every energy resource price to an international benchmark, if truly valid, would require that labour, as an exportable resource (especially in Asia), should be priced in developing countries at their international equivalents. If this were the case, the exportable, highly qualified secretaries in the Philippines would have to be paid according to the foreign scale for persons of equal quality and skill. International organizations with Asian bases (e.g., the United Nations and the Asian Development Bank) would especially have to do so. Managers, skilled workers, nurses, and doctors — all exportable — would also have to be priced at the same rate.[22] If this approach were followed, the international distribution of production to take advantage of differing resource endowments would not operate and the burden of inflation would be even more severe than it already is in developing countries.[23] This example easily demonstrates the fallacy of the argument for pegging the price of every energy resource at its international benchmark.[24]

Three other points are relevant to this issue. Shadow prices are highly judgemental in nature and therefore basically indeterminate *a priori*. They depend on value judgements by the government in determining the weights to be assigned to future consumption relative to present consumption. The weight components could be growth targets, employment levels, security of supply, and the corresponding trade-offs. Any change in the relative objectives would require a change in the estimated shadow price. Moreover, pegging a resource price to import levels would make that price sensitive to fluctuations in exchange rates, in addition to changes in world price levels and in domestic conditions.

Furthermore, for a country attempting to promote a shift away from liquid fuels, most of which are imported, a social cost may be attached to the drain such imports impose on the foreign exchange of the country and a social benefit to being able to use indigenous resources at prices possibly lower than the imported counterpart. (In fact, as already noted, it may be appropriate to address the actual scarcity problem in this case, i.e., it is foreign exchange that is scarce, not energy resources *per se*. Thus, the distortion of the energy resource price may not be the true issue, and therefore not the problem that needs to be corrected. What has to be addressed is a reduction in the demand for foreign exchange, and surely energy imports are not the only components of foreign exchange demand. Table 6.4 summarizes price distortions in different economic sectors in the 1970s in selected Asian developing countries.)

Finally, a social long-term value may also be attached to a country's diversification away from liquid fuels where import dependence is high and the stability of supply sources is uncertain (i.e., supply interruptions are always possible, and therefore disruptions of ongoing economic programs a likelihood). The long-term costs associated with such dependence and their role in shadow pricing were already pointed out earlier.

Parameters for pricing an indigenous resource

Where, for political or social reasons, it is necessary for a government to determine the appropriate price for its energy resources, parameter guidelines for pricing may be useful. At least four types of opportunity costs must be considered as starting points in assessing the appropriate rate of discount in estimating the present value of the resource:

1. For a non-oil resource, the cost of not using the resource when it is substitutable with oil, in terms of foreign exchange lost for use in economic development programmes.
2. The present value of revenues that could be gained from alternative uses (e.g., export of gas as LNG or by pipeline to adjacent countries, or conversion into chemicals such as urea, methanol, etc.).
3. The replacement costs of alternative sources that must be utilized currently and over the period of supply suggested by the stock of the resource.
4. The net revenues that would be generated by a private investor in another profit-making venture.

Thus, four parameters may be suggested for the price:

1. At the maximum, the price should provide the incentive to switch to the domestic fuel, providing for its inherent characteristics.
2. The price should not be lower than the price it could command in alternative uses, *ceteris paribus.*
3. At the minimum, the price should cover the true economic costs of production (including both private returns and all social costs), such that private investors or responsible government agencies will be encouraged not only to develop non-producing areas and explore for new ones but also to employ advanced technology to improve production and recovery efficiency.
4. The price should be high enough to provide some revenues to the government whether or not the resource is privately or publicly developed.

The foregoing implies that the resource price should basically be supply-cost based, and that it should be allowed to rise at a more natural rate according to demand-supply pressures, than that which official pegging to some benchmark price permits. Resource theory suggests that net price will rise, more or less, with the real interest rate. If the "administered" price starts at the "wrong" level (i.e., above or below the true economic price), distortion will persist. Furthermore, the resource could then be depleted more quickly or more slowly than efficiency criteria would suggest. Where the unfettered price may — even with taxation — include some economic rent, because it reflects a scarcity element, then the government may consider the employment of a resource rent tax.

The above remarks are not intended to detract from the earlier arguments

172 CRITERIA FOR ENERGY PRICING POLICY

about the administrative complexity and the potential societal costs of judgemental errors in the process of tinkering with the pricing system.

SUMMARY

In summary, in pricing its indigenous resources, a government that decides to raise prices above the direct supply-related costs to attain conservation targets or reduce foreign exchange costs needs to do so with care. The burden of proof will be on the government to show that demand-supply forces in the market are not working and that shadow pricing actually moves the economy to a relatively more efficient frontier. By definition, the amount of knowledge required for accuracy in the choice and administration of the correct shadow price is beyond that normally available even in economies with sophisticated data bases, let alone in developing countries. This paper is thus a plea for more humility and care on the part of energy planners in approaching the issue of shadow pricing.

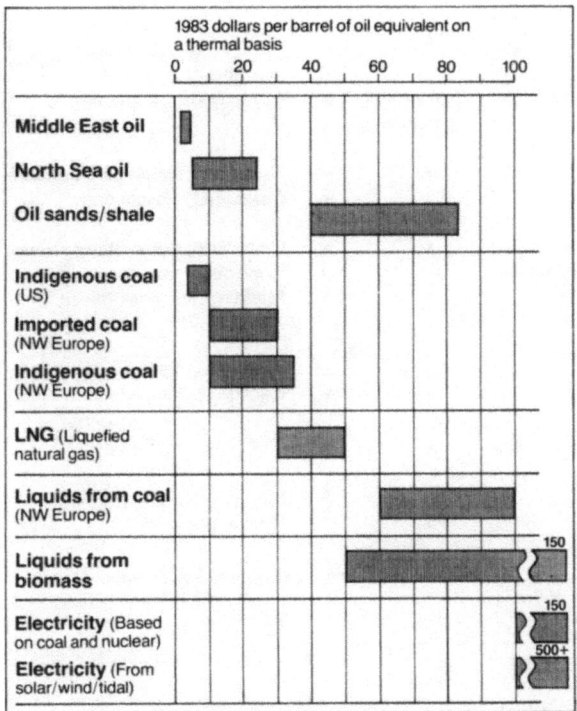

Figure 6.1 Comparative costs for alternative energy. Source: corporate records.

Table 6.1 Summary of domestic energy resource pricing policies, selected countries, as of 1981

	Energy resource	Producer		Price levels	
		Government	*Private*	*Existing*	*Proposed*
South Asia					
Bangladesh	Natural gas	x	—	Controlled; below international; economic.	Peg to world oil.
	Coal	Not produced as of reporting date.		Controlled; uneconomic.	—
India	Oil	x	—	Controlled; below international.[a]	—
	Coal	x	—	Controlled; uneconomic.	—
Pakistan	Oil	x	x	Controlled; below international.[a]	International.
	Natural gas	—	x	Controlled; uneconomic.[b]	—
	Coal	x	x	Market price; constrained.[c]	—
South-East Asia					
Burma	Oil	x	—	Controlled; below international.[a]	—
Indonesia	Oil	x	x	Supply costs (tax) (−subsidies).	International.
	Natural gas	x	x	Supply costs covered.	—
Philippines	Oil	—	x	Market price (international).	—
	Coal	—	x	Market price.	Peg to world oil/coal.
Thailand	Natural gas	—	x	Controlled by contract.	Peg to world oil.
	Coal	—	x	Supply costs.	Peg to world oil/coal.

Source: Siddayao (1983c).

[a] No good information on whether economic costs are covered.
[b] Not sufficiently attractive investments.
[c] Poor capital market has prevented investment in advanced techniques.

Table 6.2 Investment needs in the energy sector for commercial energy development in Asian oil-importing developing countries or areas, 1985–90 (Millions of current US dollars)[a]

Country or area	Coal	Natural gas	Oil	Electricity	Total
A. Low-income					
Bangladesh	—	1,091	43	1,969	3,103
Burma[b]		8	537	498	1,043
Nepal	—	—	—	544	544
Pakistan	96	1,867	10,651	5,355	17,969
Sri Lanka	—	—	43	721	764
Subtotal	96	2,966	11,274	9,087	23,423
B. High- and middle-income					
Hong Kong	—	—	—	4,040	4,040
Korea, Republic of	—	—	—	23,620	23,620
Philippines	577	—	4,084	8,421	13,082
Singapore	—	—	—	3,749	3,749
Thailand	351	1,016	87	7,735	9,189
Subtotal	928	1,016	4,171	47,565	53,680
TOTAL	1,024	3,982	15,445	56,652	77,103

Source: Table 9.12, Asian Development Bank (1982).

[a] For some countries, some exploration expenditures are projected, even though production is not expected by 1990.
[b] Burma, although self-sufficient, has been included because it is not a major net exporter and yet has vast energy resource potential.

Current prices were computed by the ADB using the midpoint inflation rates projected for 1980–1985 and 1985–1990.

Table 6.3 Asian developing countries: inflation in the 1970s compared with that in the 1960s

	Annual rate 1970–80 (percent)	Acceleration over 1960–70 (times)	Distortion classification[a]
South Asia			
Bangladesh	16.9	4.6	H
India	8.5	1.2	L
Pakistan	13.5	4.1	M
Sri Lanka	12.6	7.0	M
East-Asia			
Indonesia	20.5	—	M
Korea, Republic of	19.8	1.1	M
Malaysia	7.5	—	L
Philippines	13.2	2.3	L
Thailand	9.9	5.5	L

Source: Based on data in *World Development Report 1982* (Washington, D.C., 1982) as presented in Table 13, Agarwala (1983).

[a] Unless otherwise explained in Agarwala (1983), distortion is high (H) where the inflation rate is greater than 15 percent a year and acceleration is greater than 4 times; low (L) where the inflation rate is less than 15 percent a year and acceleration less than 4 times; medium (M) in other cases.

Table 6.4 Asian developing countries: price distortions in the 1970s

	Foreign exchange pricing			Pricing factor		Product pricing	
	Exchange rate	Production of manufacturing	Production or taxation of agriculture	Capital	Labour	Power tariff	Inflation
South Asia							
Bangladesh	L	H	M	H	H	H	H
India	L	H	M	M	M	M	L
Pakistan	M	H	M	M	H	M	M
Sri Lanka	L	M	L	M	H	M	M
East Asia							
Indonesia	M	M	L	M	L	H	M
Korea, Republic of	L	M	H	M	L	L	M
Malaysia	L	L	M	M	M	M	L
Philippines	L	M	M	M	L	M	L
Thailand	L	M	L	L	L	H	L

Source: Adapted from Agarwala (1983), Figure 1.

Legend: H = high distortion
M = medium distortion
L = low distortion.

NOTES

[1] See Siddayao (1978, 1980).

[2] The second approach is supported by opinions expressed in various reports of international organisations. (See, for example, internal country reports on Thailand, Bangladesh, and Pakistan. Other unpublished documents also support this view.) Munasinghe (1980, pp. 8, 9, and Chapter 1 in this volume) also suggests that the marginal opportunity costs of supply for fuels that are substitutes for tradable items like crude oil should be shadow priced. The shadow price would be the international or border prices of the tradables with adjustments for internal costs. He acknowledges, however, that coal and natural gas may or may not be tradables (Munasinghe, 1980, p. 9).

[3] See, for example, the studies summarized in Ranada (1982). More recent reviews of the United States case are found in Sweeney (1984) and Gordon (1984). Various in-country studies in Asia and elsewhere abound. See the references cited in Siddayao (1983).

[4] See Newbery's discussion on "market failure" in Chapter 3, this volume.

[5] See Hughes and Singh (1978) and Hughes (1975).

[6] Bruno (1972).

[7] See Squire and van der Tak (1975), p. 26.

[8] Little and Mirrlees (1969), p. 161.

[9] Squire and van der Tak, p. 32 ff.

[10] The term "user cost" used with reference to depletable resources refers to the intertemporal opportunity cost associated with producing the resource today as opposed to producing it at some future date. See the discussion in Chapter VII in Siddayao (1983) on rent-sharing. It is what may be referred to as the present value of the stock of resources which is sacrificed by producing the resource today.

[11] See Chapter 3 in this volume, by Newbery, for an expansion on the issue of "market failures".

[12] Squire and van der Tak, p. 36.

[13] See Hotelling (1931).

[14] See Hotelling (1931), Gordon (1967), and Solow (1974).

[15] Nordhaus (1973) and Dasgupta and Stiglitz (1976).

[16] See discussions on "political risk" and references cited in Siddayao (1980).
[17] Details on the notion of measuring resource rents are given in Hughes (1975) and Garnaut and Ross (1975).
[18] See Siddayao (1980), Johnson (1981), Garnaut and Ross (1983), and Siddayao (1984).
[19] See Siddayao (1975a, b) and relevant references cited therein.
[20] See Lipsey and Lancaster (1956).
[21] Chapter II in Siddayao (1983) discusses this in more detail. Other studies may argue that this effect is small, but no model can honestly claim that it has captured reality.
[22] Only immigration rules limit their outflow to industrialized or capital-surplus countries. Those who argue that labour is not a traded commodity might want to reconsider that point. The labour force of a country is one of its basic economic resources. Tremendous technological and institutional changes have taken place over time that have placed labour services in the same category as commodities, even if the act of trading is conducted in subtler and different forms.
[23] Trade specialists may argue against the treatment of energy in this analysis as a factor of production. Such critics are referred to the earlier discussions both in this chapter and in Chapter VII of Siddayao (1983) on the energy theory of value that has dominated energy policy analysis and to models that treat energy as a separate variable input in the production function. If energy is to be accorded such status in production function analysis, one can argue that labour and capital need not be discriminated against in pricing analysis.
[24] One might argue that the shadow wage rate of these skilled workers or professionals is zero in an underdeveloped country because of the high unemployment rate. This is an incorrect argument, because it is inappropriate to treat these units as homogeneous. One could view each worker's skills as a product, such that there are several sets of heterogeneous products with different demand/supply conditions attached to them.

REFERENCES

Agarwala, R. (1983). *Price Distortions and Growth in Developing Countries*. Staff Working Papers No. 575. Washington, D.C.: The World Bank.

Asian Development Bank (1982). *Asian Energy Problems: A Regional Energy Survey*. New York: Praeger Publishers.

Bruno, M. (1972). "Domestic resource costs and effective protection: clarification and synthesis." *Journal of Political Economy*, Vol. 80, No. 1 (January/February).

Dasgupta, P. S. and J. E. Stiglitz (1976). *Uncertainty and the Rate of Extraction Under Alternative Institutional Arrangements*. Technical Report No. 179, Institute for Mathematical Studies in the Social Sciences. Stanford: Stanford University.

Garnaut, R. and A. Clunies Ross (1975). "Uncertainty, risk aversion and the taxing of natural resource projects." *Economic Journal*, Vol. 85, No. 338 (June), pp. 272–287.

Garnaut, R. and A. Clunies Ross (1983). *Taxation of Mineral Rents*. Oxford: Clarendon Press.

Gordon, R. L. (1967). "A reinterpretation of the pure theory of exhaustion." *Journal of Political Economy*, Vol. 75, No. 3 (June), pp. 274–286.

Gordon, R. J. (1984). "Supply shocks and monetary policy revisited." *American Economic Review*, Vol. 74, No. 2 (May), pp. 38–43.

Hotelling, H. (1931). "The economics of exhaustible resources." *Journal of Political Economy*, Vol. 39.

Hughes, H. (1975). "Economic rents, the distribution of gains from mineral exploitation, and mineral development policy." *World Development*, Vol. 3, Nos. 11 and 12.

Hughes, H. and S. Singh (1978). "Economic rent: Incidence in selected metals and minerals." *Resources Policy*, Vol. 4, No. 2 (June), pp. 135–145.

Johnson, C. J. (1981). "Considerations in establishing an effective production sharing type tax regime for petroleum." *Resources Policy*, Vol. 7, No. 2 (June).

Lipsey, R. G. and K. Lancaster (1956). "The general theory of second best." *Review of Economic Studies*, Vol. 24, No. 63, pp. 11–32.

Little, I. M. D. and J. A. Mirrlees (1969). *Social Cost Benefit Analysis*, Vol. II of *Manual of Industrial Project Analysis*. Paris: Development Centre of the Organisation for Economic Co-operation and Development.

Munasinghe, M. (1980). "An integrated framework for energy pricing in developing countries." *Energy Journal*, Vol. 1, No. 3 (July), pp. 1–30.

Newbery, D. M. G. (1981). The Taxation of Oil Consumption. Report commissioned by the Policy Review Unit, British Petroleum. London: 20 July 1981. Manuscript.

Nordhaus, W. D. (1973). "The allocation of energy resources." *Brookings Papers on Economic Activities*, No. 3.

Ranada, J. G. (1982). *The Economic Impact of Rising Oil Prices: A Survey of Theory and Methodology*. Resource Systems Institute Working Paper Series WP-82–15. Honolulu, Hawaii: The East-West Center.

Siddayao, C. M. (1975a). "Natural gas problems of the United States: Causes and alternatives for the future." Report prepared for a National Science Foundation funded project of the George Washington University's Energy Policy Research Project.

Siddayao, C. M. (1975b). *The Role of Field Price Regulation of Natural Gas in Its Use for Electricity Generation*. Ph.D. Dissertation. George Washington University. Ann Arbor, Michigan: University Microfilms.

Siddayao, C. M. (1978). *The Off-shore Petroleum Resources of South-East Asia: Some Potential Conflicts and Related Economic Factors*. Kuala Lumpur: Oxford University Press.

Siddayao, C. M. (1980). *The Supply of Petroleum Reserves in South-East Asia: Economic Implications of Evolving Property Rights Arrangements*. Kuala Lumpur: Oxford University Press.

Siddayao, C. M. (1981a). *Fossil Fuel Pricing Policies in the Asia-Pacific Region: A Preliminary Assessment of Some Allocative Implications*. Expanded version of report prepared for the Asian Development Bank 1980 Regional Energy Survey issued in the Resource Systems Working Paper Series as WP-81–3. Honolulu: The East-West Center.

Siddayao, C. M. (1981b). *Pricing of Fossil Fuels in Asia: Allocative Implications*. Paper presented at the International Atlantic Economic Conference, London, 1981, and issued in the Resource Systems Working Paper Series as WP-81–11. Honolulu: The East-West Center.

Siddayao, C. M. (1983). *Oil Prices, Balance of Payments, and Asia's Growth: Interaction of Public and Private Sector Responses to Energy Demand/Supply Issues*. Copyrighted manuscript.

Siddayao, C. M. (1984). "Book review of *Taxation of Mineral Rents* by Ross Garnaut and Anthony Clunies Ross." *Journal of Energy and Development*.

Solow, R. M. (1974). "The economics of resources or the resources of economics." *American Economic Review*, Vol. 64, No. 3, pp. 1–21.

Squire, L. and H. G. van der Tak (1975). *Economic Analysis of Projects*. Baltimore: The Johns Hopkins University Press (for the World Bank).

Sweeney, J. L. (1984). "The response of energy demand to higher prices: What have we learned?" *American Economic Review*, Vol. 74, No. 2 (May), pp. 31–37.

Warr, P. G. (1980). "Shadow pricing rules for non-traded commodities." *Oxford Economic Papers*, Vol. 34, No. 2, pp. 305–325.

REPORT OF THE REGIONAL ENERGY DEVELOPMENT PROGRAMME ESCAP/ILO/EEC/EWC/IRDC ENERGY PRICING POLICY WORKSHOP

I. ORGANIZATION OF THE MEETING

A. Background

1. The Workshop co-financed by the United Nations Development Programme (UNDP) under the regional energy development programme (REDP) with the Commission of the European Communities (EEC) and the International Development Research Centre (IDRC) with substantial contributions in kind by the International Labour Organisation (ILO) and the East-West Center (EWC) had as its main objective to outline certain in-country pricing policy implementation studies to be carried out under the regional energy development programme. A subsidiary objective was to bring together worldwide expertise in energy pricing in order to synthesize the best available information for subsequent publication for wider dissemination.

B. Organization of the meeting and attendance

2. The meeting was organized by the Economic and Social Commission for Asia and the Pacific (ESCAP) with the collaboration of the East-West Center, at Bangkok, from 8-11 May 1984.

3. The country experts were asked to bring to the Workshop their pricing policy implementation problems to be formulated as projects for implementation with the help of the experts present.

4. Experts from China, India, Indonesia, Nepal, the Philippines, Sri Lanka and Thailand participated. In addition, observers from the United Nations Department of Technical Co-operation for Development and the Asian Institute of Technology were present as were representatives of the sponsoring organizations.

C. Opening of the meeting

5. The meeting was inaugurated by the Executive Secretary of ESCAP and the Assistant Director-General of ILO. In his inaugural address the Executive Secretary emphasized the importance of the inter-organizational co-operation orchestrated by ESCAP under the regional energy development programme and wished the Workshop well. The Assistant Director-General expressed his appreciation of ESCAP's efforts in orchestrating such a co-operative project and outlined ILO's energy development activities.

D. Election of officers

6. Mr. G. Makasiar of the Philippines was elected Chairman, Mr. N. Mathanagopalan of Sri Lanka and Mr. Huang Zhijie of China as Vice-Chairmen and Mr. A.K. Mago of India as Rapporteur. Later, Mr. S.N. Sharma of Nepal was appointed as a third Vice-Chairman by the Chairman so as to have a Vice-Chairman chair each of the three planned working groups.

E. Adoption of the agenda

7. The meeting adopted the following agenda.
    1. Opening of the Workshop
    2. Election of officers
    3. Adoption of the agenda
    4. General criteria for energy pricing policy
    5. Pricing policy in practice
    6. Country studies taking account of social welfare criteria
    7. Design of follow-up studies
    8. Adoption of the report
8. It was agreed that the Workshop would concentrate on consultants' papers on the first day (agenda items 4 and 5), and country pricing problems on the second day (agenda item 6), with three working groups formulating follow-up studies on the third day (agenda item 7). General conclusions and recommendations would be summarized after the presentation of the working group results, on the fourth day.

9. EWC, EEC and IDRC briefly outlined their expectations from the meeting, with EWC describing the history of the Workshop from its inception at the 1981 Session of the Committee on Natural Resources. The Chairman then called on the EWC participant to introduce agenda item 4.

II. PROCEEDINGS

A. Agenda item 4

10. Consultants' papers by Messrs Kumar, Newbery and Bhatia were presented, with subsequent discussions outlining general criteria for pricing policy with some illustrative examples.

11. The paper by M.S. Kumar, "Socio-economic goals in energy pricing policy: a framework for analysis", provided a framework for analysing the role pricing of energy could play in meeting the socio-economic goals of developing Asian countries. It extended the traditional approach to pricing based on marginal cost and financing considerations and examined how pricing policies could affect goals relating to equity, employment, industrialization, export competitiveness and rural development. In analysing issues of equity, the paper emphasized that it was important to consider the absolute change in the real income of the poorest groups as well as the change in the whole distribution of income. It next emphasized that in the medium- and long-term there could be a considerable effect on employment as a result of changes in the techniques of production in industry and in agriculture and changes in the product mix. The paper also analysed the way in which growth of industry as a whole, and of the subsectors within it, were likely to be affected by energy price changes. The paper then identified the availability of foreign exchange as a major constraint on the developmental process and examined how changes in prices would affect that constraint, both by affecting the efficiency with which energy was utilized, and by affecting costs and export competitiveness. The paper also examined some issues relating to the use of traditional fuels and the role prices could play in leading to substitution between those and commercial fuels. The paper ended by emphasizing the need for detailed empirical estimates about the response of different sectors of the economy to energy price changes.

12. In the paper by David Newbery, "Efficiency and equity criteria in energy pricing with practical application to developing countries in Asia", a comparison was made of policy responses in the United States of America and the United Kingdom of Great Britain and Northern Ireland to the 1974 oil price rise. That comparison suggested that a potent tax system allowed the goals of equity and efficiency to be pursued separately, but its use might have been constrained by the redistributive effects of the price rise. In the United States, complex and administratively costly measures were required and the process of adjustment to an efficient price regime was slow. The United Kingdom's experience suggested that the redistributive effects of the oil price rise were small and on balance probably beneficial. In less developed countries the tax system was less potent, but for most fuels the problem was simplified by the fact that they were either consumed by producers or final consumers. In the former case prices should be guided by efficiency criteria, and only in the latter case was equity a problem. The main exception was kerosene which was highly substitutable for diesel. A study of the effect of bringing kerosene prices up to their efficient level in Thailand suggested that the redistributive impact was small and random, and that therefore subsidizing kerosene was an inefficient way of redistributing income to the poor. The other main source of conflict between equity and efficiency was likely to occur for rural electricity,

where the gap between marginal and average cost was very wide, and where electricity was consumed by relatively richer rural households. The paper set out a methodology for calculating the distributional impact of price changes on consumers based on a significantly different pricing rule than that of conventional mark-up pricing models, and which identified the impact on factor incomes, using household budget and input-output data.

13. The paper by Ramesh Bhatia, "Energy pricing in developing countries: role of prices in investment allocation and consumer choices", pointed out that energy prices played an important role in the allocation of investments and consumer choices. In order to meet the objectives of equity and control of inflation, Governments might fix prices which were so low that the producers did not make adequate profits for investing in new facilities and modernization schemes. That resulted in shortages of supplies which, in turn, adversely affected economic development. Examples from the coal and electricity sectors in India were given to show how low prices adversely affected investment and output in those sectors. It was also pointed out that due to overall shortages, poor people did not actually get energy supplies at the administered prices. The paper also showed how low consumer prices for some of the energy sources (e.g. kerosene, diesel and electricity in rural areas) distorted the choices of consumers, resulted in investments in "back-up" systems, and acted as disincentives for energy conservation and development of renewable sources. Those points were illustrated by studies on the transport and irrigation sectors in India and domestic/transport sectors in Sri Lanka and Bangladesh. It was suggested that the objectives of equity and efficiency could be reconciled by (a) direct subsidies for target groups; (b) extending the scope of investments such as rural electrification; (c) subsidies on costs of equipment rather than on fuels; and (d) explicit subsidies on renewable energy technologies. It was emphasized that there was need for an integrated energy pricing policy which included analyses of prices of all fuels, prices of other factors of production, and levels of taxes and subsidies on energy and non-energy inputs.

B. Agenda Item 5

14. Agenda item 5 was introduced by the ILO participant and four papers on practical implementation aspects of energy pricing policy followed.

15. The paper by Corazon Siddayao, "Shadow pricing indigenous energy: its complexity and implications", addressed the issue of using the average c.i.f. border price of imported energy to shadow-price domestic products and indigenous resources. Arguments presented included:

   (a) Efficiency and effectiveness criteria depended on objectives and the socio-economic environment;
   (b) Because of the above, generally valid recommendations on shadow pricing would require extensive data that were not usually available to policy-makers;

(c) "Rule of thumb" policies were likely to lead to persistent distortions and misallocation of economic resources that were hard and time-consuming to correct;

(d) The policy analyst was thus faced with an eclectic, case-by-case analysis of each energy pricing problem, and had also to keep in mind trade-offs between the interests of present and the needs of future generations.

16. The paper by Gunter Schramm, "Operationalizing efficiency criteria in energy pricing policy", consisted of three parts. The first defined the plural objectives of pricing policies consisting of efficiency, equity and financial considerations. The second considered appropriate approaches to calculating the various levels of efficiency prices which could then serve as basic measuring devices for analysing the costs of all other objectives. The third applied the principles developed to a series of actual case studies which were used to illustrate the application of the criteria, develop appropriate methodologies for calculating costs and prices, and show the results under a variety of real world situations. Efficiency prices were shown to be defined by five different types of opportunity costs: the long-term marginal costs of supply, the future costs of depleting (if applicable), the net value of the resource in terms of freely traded border prices, the net value of the resource as a substitute for other energy resources net of all differences in systems costs, and the value of the resource in uses that would not occur in its absence. Case studies addressed, among other issues, the critical effects of depletion costs on alternative allocations of limited resources of natural gas, which was found to be of particular importance to a number of countries in the region. Other case studies dealt with the inimical effects and high economic costs of disregarding financial criteria in addition to economic ones in setting prices, the potential for using fuel taxation over and above economic costs to finance public transport sector expenditures, the difference between border prices and the actual economic opportunity costs of domestic petroleum resources that were subject to export quotas, the real economic cost of uniform power tariffs and discriminatory rates, and the practical problems of measuring and allocating joint costs of supplies in gas distribution systems with large numbers of residential users.

17. The paper by Sauter-Servaes, "Energy pricing policies in the European Community", first outlined trends of energy demand and supply and of energy prices in the European Community emphasizing a drop in primary energy demand (especially crude oil), an increase in indigenous energy production and a sharp divergence in the trends of prices for oil on the one hand, and coal on the other during the period 1979–1983. The objectives of the agreements on energy pricing into which Member States had entered at the Community level were summarized as (a) full cost-bearing (including long-term costs) by energy consumers, (b) elimination of differences in policies and practices which gave rise to distortions, and

(c) transparency of energy markets. The paper finally focused on the last point and described the activities and difficulties of the Commission in publicizing energy prices.

18. The paper by Mohan Munasinghe, "Energy pricing policy framework and experience in developing countries", showed how recent increases in energy supply costs led to increased emphasis on the importance of integrated national energy planning (INEP) and pricing in developing countries. It described a comprehensive framework for energy policy formulation which explicitly recognized three hierarchical levels of analysis: energy-macroeconomy, energy sector (including subsector interactions), and energy subsectors (supply and demand management). The policy instruments available for energy demand management and conservation included physical controls and technical methods which were more effective in the short-term, as well as the medium- to long-term tools like pricing, financial incentives, education and promotion. It was held that co-ordinated use of those policy tools provided the best results. Pricing policy was developed in two stages. First the strictly efficient price of energy supply was calculated based on the (shadow-priced) marginal opportunity cost, suitably adjusted (on a second-best basis) for demand-side distortions in prices of other goods and services. That price satisfied the economic efficiency objective of pricing policy. The efficient price was then systematically adjusted to meet the other objectives and constraints of pricing, including the basic needs of poor consumers, financial requirements of the energy-producing institutions, customer comprehension and simplicity of price structure, price stability, and other special considerations. Some recent experience in the pricing of electricity and petroleum products in Asian developing countries was discussed. It was maintained that energy conservation was an aspect of demand management which should be pursued on the basis of economic viability rather than purely technical considerations. The basic criterion was that the benefits represented by fuel savings should exceed the costs of implementing the conservation measure plus the foregone benefits of reduced energy consumption. A Government's decision to support a conservation project must be determined on the basis of economic opportunity costs. Once that criterion was met, the authorities could adjust market prices, if necessary, to induce private individuals to adopt the measures. A case study illustrating the co-ordinated use of price and non-price tools for energy conservation was presented.

C. General discussion

19. In the general discussion after the presentations it was emphasized that a considerable degree of consensus had been achieved among the experts concerning the necessary steps and procedures to be followed for a rational pricing policy. Some misgivings were expressed with respect to that consensus, for although efficiency questions had been discussed at

great length, the central role of socio-economic factors had been touched upon only in terms of marginal aberrations.

D. Agenda item 6

20. The agenda item was introduced by the secretariat, with the announcement that the country presentation on pricing practices would form the basis for discussion by working groups in formulating pricing policy implementation projects. Summaries of the presentations were as supplied by the experts.

1. China

21. It was maintained that price in China did not play the critical role in supply and demand adjustment that it did in other countries. However, the existing energy price did influence energy exploitation, conservation and rational utilization. For instance, a low profit rate in the coal industry held up exploitation. Low energy prices hindered energy conservation. Nonrational utilization of energy caused by unreasonable price ratios between various kinds of energy and low energy prices hindered utilization of renewable energy. Thus, although China was an energy self-sufficient country, it still had a need to study energy pricing policy and to solve related problems.

2. Indonesia

22. Indonesia, through the state-owned electricity corporation (PLN), with the World Bank initiative, had moved in the direction of long-run marginal cost pricing, beginning in 1979. PLN's tariff was uniform across the country. The recent PLN-1984 tariff was averaged at Rp 98/kWh (US $0.09/kWh), which was among the lowest among the members of the Association of South-East Asian Nations. The basic tariff consisted of 17 categories for residential, commercial and industrial users. There were cross-subsidies; for example, the rich subsidized the poor. The growth rate of the use of electricity was high, about 20 percent during the 1970s, due to the increased area of electrification. The real price of electricity in the industrial/business sector was actually declining at an average of 3.14 percent, despite the fact that the real regional income had increased significantly by an average of 10.74 percent per annum. The real price of electricity in the residential sector during the 1970s had increased at an annual rate of 4.17 percent as compared with the 8.55 percent increase in real per capita income. It was considered that by some diversification in the generation of electricity, upward pressure on the price should somehow be reduced. PLN had moved in that direction, for example, by the use of geothermal power plants (1983), natural gas power plants (1983) and the coal-fired steam power plant expected to be commissioned in 1984. Efficiency in producing electricity should be obtained by the development of a load dispatch centre for the Java system. In order to allocate resources efficiently, it was believed that price should reflect the opportunity cost of providing electricity. If resource allocation was a long-term objective, then

governmental policy should be gradually geared toward an efficient pricing scheme.

3. Nepal

23. The energy consumption pattern in Nepal mirrored the rural agricultural nature of its economy and the small size of its modern industrial and commercial sectors. Currently the major infrastructure requirements lay in the energy sector. The exploitation of the forests for fuelwood had created the spectre of ecological disaster; large-scale afforestation and reforestation programmes were thus essential, along with ingenuity in satisfying rural energy needs. Developing the abundant indigenous hydropower resources was another formidable task. Electricity was a particularly important energy form. Hydropower was the country's only known source of commercial energy and the potential for development was considerable. Rural energy, which was primarily fuelwood, was not supplied through a market system but rather was collected by the user on a day-to-day basis for immediate consumption. Urban consumers typically had a wider variety of energy forms and an established market system through which to purchase them. The pricing policy of the Nepal Fuel Corporation, which was the major supplier of fuelwood to urban areas, was to set prices to cover production costs. Those prices did not include resource costs. All petroleum products were imported and their retail prices were reflective of international prices. The tariff history of electricity was a reflection of the type of expansion of the system. It was unlikely that any government action to impose taxes or a pricing mechanism on rural energy would meet with any success. A programme to install improved stoves, free of charge, and to train villagers in basic forest management would have far better returns. Electricity tariffs were well below both those required for a reasonable rate of return on assets and those based on long-term marginal costs.

4. Thailand

24. Total energy consumption in 1982 was $18,120.54 \times 10^6$ litres of crude oil equivalent. It was classified into petroleum products 60 percent, natural gas 7.1 percent, hydroelectric resources 7.2 percent, bagasse 6.4 percent, coal and lignite 4.2 percent, fuelwood and charcoal 14.3 percent and paddy husk 0.8 percent. Thailand relied on foreign sources for up to 60 percent of its energy supply, especially crude oil and petroleum products. The energy supply from indigenous sources in 1982 was classified into hydroelectric 14 percent, lignite 8.7 percent, fuelwood and charcoal 33.6 percent, paddy husk 8.0 percent, bagasse 16.0 percent, petroleum products 2.9 percent and natural gas 16.8 percent. Electricity and petroleum consumption in Thailand was classified by economic sectors as agriculture 9.1 percent, manufacturing industry 29.4 percent, transportation and communication 33.4 percent, construction 0.9 percent and commerce, service, and other activity 27.2 percent. Thailand had faced problems in maintaining its economic progress on account of higher energy costs since 1973.

Before 1974, the country's energy demand expanded at a rate of 13 percent per year. After the big oil price increase, the country's energy demand increased at a rate of 7 percent per year. This was still high because the Government had been trying to insulate domestic energy consumers from the effect of the external oil price explosion. The Government substantially adjusted domestic prices in 1980 and 1981. Pricing of various petroleum products was still distorted and inappropriate, due to differences in tax rates or direct subsidy. Domestic energy pricing structure, problems of oil supply and stockpiles, and oil refinery capacity problems were closely interrelated. The most fundamental problem facing Thailand in economic development was the rising price of imported oil. Thailand spent 37 percent of the foreign exchange it earned from exports on importing oil, totalling approximately Baht 58,799 million in 1982. With strategies, measures and appropriate action on energy pricing policy, Thailand would be in a better position economically, financially and socially.

5. India
25. The energy scene in India was briefly described. India had modest energy resources considering its population. Energy policy in India had laid stress on accelerated exploitation of domestic energy resources, management of demand, energy conservation, development of renewable sources of energy etc. The mechanism for energy pricing in different subsectors was explained. Pricing in the energy sector was largely based on cost and "retention prices." The current pricing policy was not based on the long-term marginal cost. There was a need to develop an integrated system of energy pricing which would help in an economically efficient allocation of resources intersectorally in the energy subsectors, help raise additional resources for the expansion of the energy sector, and take into account the need to meet the minimum energy requirements of low-income groups at reasonable prices.

6. Sri Lanka
26. The Ceylon Electricity Board (CEB) was responsible for the generation, transmission and distribution of electricity except for the 218 local authorities (such as municipalities and urban councils) which bought electricity in bulk from CEB and distributed and sold it to the consumers in their respective areas. In the early 1970s the annual growth rate of total sales to consumers was in the range of 3 to 4 percent. In the year 1977 due to a change in government policy (in that an open economy system was introduced) a sudden upsurge in development was created, especially in the industrial and commercial sectors. That caused a sudden increase in the annual sales in the domestic (due to large purchases of electrical appliances and their usage), industrial, and commercial sectors, and the annual growth rate of electricity demand in the early 1980s was three times what it had been in the early 1970s. In order to cater to that rapid increase in demand CEB had to make large investments in generation, transmission and distribution expansion programmes. In order to implement those

programmes CEB had to borrow from lending agencies such as the World Bank and the Asian Development Bank. Loan agreements were signed in the early 1980s with those agencies incorporating such requirements as that CEB should earn a minimum 8 percent return on net revalued fixed assets, that a debt service cover of 1.25 times should be maintained, and that CEB should carry out a study of the long-run marginal cost of electricity and revise its tariffs in keeping with that study. With assistance, CEB carried out the study in 1981 and revised its tariffs in mid-1982. As an example, the maximum demand charge rate was increased five-fold. That brought home to consumers the message that "Electricity is expensive: use it, but don't waste it." The problems that CEB was currently examining concerned:

(a) The realistic life of a line block of units in the domestic sector and at what rate it should be sold;

(b) The number of units that should be allocated to each block of units in the domestic sector and at what rates. In order to determine that, CEB was about to launch a statistical survey of its domestic consumers;

(c) Local authorities had argued that the price at which CEB sold them electricity in bulk made it not viable for them to sell to their consumers at a reasonable price and make the profit needed to maintain and improve their systems;

(d) The system losses which were at a level of 20 percent in the late 1970s were currently about 17 or 18 percent and needed further improvement. The question was whether the entire improvement cost should be borne by CEB or whether part of it should be borne by consumers.

7. Philippines

27. The history of energy pricing practices in the Philippines reflected a mixture of both open market competition and administered pricing. In general, international prices were accepted for all imported fuel forms. For locally-produced fuels, treatment varied according to type. Border prices were used as reference for tradable fuels, adjusted mainly for quality differences. Coal, firewood, and charcoal were freely traded in the domestic market. More problematic in terms of valuation were site-specific resources, such as hydro and geothermal steam. Prices charged by different agents in the electricity chain were regulated by respective regulatory boards. Bulk and retail power rates were designed to achieve targeted revenues which were pre-computed to yield "acceptable" rates of return on assets. Industrial petroleum product prices and taxes followed a socialized scheme (fuels identified with affluent consumers carried more of the burden), with the target result that composite refinery revenues covered all government-recognized or allowable costs/expenses. Some of the more important issues confronted in energy pricing decisions included economic as well as politico-social considerations that were equally valid, such as:

(a) The amount by which the Government should bite into the economic surplus (of either producers or consumers) from the exploitation of patrimonial resources and the allocation of that amount between the national and local levels;

(b) The timing of implementation (gradualism or automaticity) especially where prices had to be increased;

(c) The allocation of the burden of cross-subsidies and for how long at a time, if subsidies were a political necessity (on a temporary or permanent basis); and the size of that constituency compared with that of those who benefited from the subsidy.

E. Agenda item 7

28. The Workshop broke up into the following working groups:

(a) Sri Lanka (Chair) Indonesia	Experts	*Schramm* Munasinghe Kumar
(b) Nepal (Chair) India Thailand	Experts	*Newbery* Bhatia Amjad Goldsmith
(c) China (Chair) Philippines	Experts	*Siddayao* Desai Sauter-Servaes Dewulf

29. The project summaries formulated by the working groups were as follows:

China

Organizations involved (tentative):

(a) Pricing Centre (under Pricing Commission)

(b) Energy Research Institute of State Economic Commission

(c) Economic Research Institute (under Chinese Academy of Social Science)

What needed to be done (what was the problem?); Background and justification (*Why* was the problem important?):

(a) What energy pricing policy could limit the demand of certain energy forms?

(b) What energy pricing policy could promote the utilization of alternative energy?

(c) What energy pricing policy could promote energy conservation?

(d) What energy pricing policy could promote the rational utilization of energy?

(e) What energy pricing policy could promote economic development and decrease the impact on the living standard of low-income households?

After the rise in world-wide energy prices in the last decade, most of the oil-importing developing countries faced balance-of-payments deficits and rising rates of inflation. A reasonable energy price would limit the demand for certain forms of energy and promote the utilization of alternative forms of energy. For the energy self-sufficient developing countries a reasonable energy price would promote energy conservation and rational utilization of energy. Yet there was no convincing theory and method to formulate reasonable energy prices. Therefore, energy pricing policy was necessarily the subject of study, and appropriate solutions were required.

How was the work-plan to be implemented?
Step-by-step description and timing:
    (a) Study the influence of energy pricing on:
        (i) Energy consumption;
        (ii) Conservation of energy;
        (iii) Equipment renewal and reconstruction;
        (iv) Exploitation and utilization of nuclear and renewable energy;
        (v) Economic growth rate;
        (vi) Rates of inflation;
        (vii) Employment etc.;
    (b) Study the theory of energy pricing policy;
    (c) Study the method of energy pricing policy;
    (d) In different kinds of countries (such as oil-importing countries, energy self-sufficient countries, and energy-exporting countries) investigate energy pricing policy and apply properly formulated methods to calculate optimum energy prices;
    (e) Give the results in appropriate reports.

How were results to be monitored and evaluated? (Criteria of success):
    (a) Evaluation by responsible organization in countries investigated;
    (b) Evaluation by experts of the workshop;
    (c) Social evaluation, after the study was published.

India I
Organizations involved (tentative):
Department of Power (Energy Policy Wing), Ministry of Energy

What needed to be done (What was the problem?); Background and justification (*Why* was the problem important?):
How could the minimum energy needs of the low-income groups in rural and urban areas best be met? Various ways like rural electrification, subsidized kerosene, social forestry, subsidized biogas plants, fuelwood burning stoves, etc. were being supported by the Government, and the aim was to identify the most cost-effective way of meeting the energy needs of low-income groups. For the country's seventh plan, which was under formulation, the question of how best to meet the minimum energy needs of the low income groups was considered a high-priority item.

How was the work-plan to be implemented?
Step-by-step description and timing:

Analysis of the available data in one region of India would have to be undertaken by a consultant. The region would be selected on the basis of availability of data, representativeness of region, level of electrification etc.

How were results to be monitored and evaluated? (Criteria of success):

The recommendations which resulted from the study would be considered by the Government for implementation.

India II
Organizations involved (tentative):

Department of Power (Energy Policy Wing), Ministry of Energy

What needed to be done (What was the problem?); Background and justification (*Why* was the problem important?):

A social cost-benefit analysis would be made of ground-water irrigation in eastern Uttar Pradesh, identifying the role of energy prices in the demand for irrigation, and the impact of irrigation on employment and income distribution. The study would indicate changes in prices of fuels, tax subsidies on energy equipment, and priorities in rural electrification needed in order to improve utilization of ground water. Uttar Pradesh had a large potential for ground-water irrigation which appeared to be seriously underutilized.

How was the work-plan to be implemented?
Step-by-step description and timing.

A consultant would need to analyse the available data and test the results by field visits.

How were results to be monitored and evaluated? (Criteria of success):

The criterion of success was that the explanation for underutilization was convincing and suggested possible policies for improving the situation if indeed it was socially profitable to increase irrigation.

Indonesia I
Organizations involved (tentative):

What needed to be done (What was the problem); Background and justification (*Why* was the problem important?):

An analysis would be undertaken of social and economic consequences of full-cost utility pricing. The utility had insufficient revenue flows to cover its general expenditures and the cost of heavy investment programmes. It depended on government allocations of budgetary funds and outside financing, both of which were inadequate to cover needs. As a consequence, operating performance was impaired, maintenance was inadequate and losses were high.

How was the work-plan to be implemented?
Step-by-step description and timing:
  Suggested focus on Java because of data constraints.

How were results to be monitored and evaluated? (Criteria of success):
  Expected results:
  (a) Calculation of needed levels of revenue to cover defined perfor-
      mance criteria;
  (b) Comparison between existing tariff levels and tariff needs;
  (c) Potential effects of raised tariffs on consumers and producers.

Indonesia II
Organizations involved (tentative):

What needed to be done (What was the problem); Background and
justification (*Why* was the problem important?):
  An analysis would be made of the impact of tariff changes on energy
choices by industry. Private generating capacity roughly equalled installed
public utility generation capacity. Higher industrial/commercial tariffs
might continue to reinforce that trend. The study was to analyse the consequences of higher tariffs and of possible policy measures (e.g. changing
fuel prices, regulation, etc.) to establish desired patterns.

How was the work-plan to be implemented?
Step-by-step description and timing:
  Establish comparative costs of:
  (a) Potential of using auto-generating plants;
  (b) New plants.
  Compare private and social costs of auto-generation.

How were results to be monitored and evaluated? (Criteria of success):
  Other implications:
  (a) Potential of using private generation for peak shaving;
  (b) Implications of private generation for utility reliability standards.

Indonesia III
Organizations involved (tentative):

What needed to be done (What was the problem?); Background and
justification (*Why* was problem important?):
  Pricing of coal. Coal was being developed as a new fuel for power generation. Coal could be imported or domestically supplied by a new state-owned
coal mine. What should be the transfer price between the latter and the
utility, and why? Should imports be allowed in addition and at what price?

How was the work-plan to be implemented?
Step-by-step description and timing:
  (a) Study import coal prices;

    (b) Evaluate costs of domestic coal;
    (c) Compare both in economic and financial terms;
    (d) Analyse effect of different coal price levels on:
       (i) Mining operation;
      (ii) Utility costs.

How were results to be monitored and evaluated? (Criteria of success):

Nepal
Organizations involved (tentative):
  Water and Energy Commission Secretariat

What needed to be done (What was the problem?); Background and justification (*Why* was problem important?):
  Electricity, fuelwood and kerosene were sold in urban areas at heavily subsidized prices which external funding agencies had recommended should be decreased. The aim was to identify the socio-economic impact of fuel price rises in urban areas, using consumer survey data. A secondary aim would be to identify alternative ways of preserving the standard of living of low-income groups by other policies.

How was work-plan to be implemented?
Step-by-step description and timing:
  A consultant would be found to analyse the data and write a report.

How were results to be monitored and evaluated? (Criteria of success):
  A successful report would identify the impact of changing prices on different categories of urban consumers, and would provide the Government with useful information for designing a new energy price structure and complementary reforms which would protect the urban poor.

The Philippines
A. Organizations involved (tentative):
  (See below under item D, No. 8.)

B. *What* needed to be done (What was the problem?)
  Rank I: Geothermal steam pricing.
  Rank II: Rural electricity pricing impact on labour-
             displacement.
  Rank III: Income-distributive impact of selected energy
             technologies.

C. Background and justification (*Why* was problem important?):
   I. Operational urgency.
   II. Long-term policy implications.
  III. Long-term policy implications.

D. How was the work-plan to be implemented?
  Step-by-step description and timing:

Activities	Study I	Study II	Study III
(1) Terms of reference defined	2 weeks	1 month	1 month
(2) ESCAP consultations } (3) Selection of consultant }	1 month	1 month	1 month
(4) Formulation of methodology proposed (by consultant)	1 month	1½ months	1 month
(5) Conduct of study	6–8 months	6 months	4 months
(6) Draft recommendations including implementation scheme	2 months	2 months	2 months
(7) Final report	1 month	1 month	1 month
(8) Organizations	Ministry of Energy National Power Commission	National Energy Agency	Ministry of Energy National Economic Development Agency

E. *How* were results to be monitored and evaluated? (Criteria of success)
   (a) Would prefer that procedure for monitoring and evaluation of implementation need not be explicit or formal; instead these should be undertaken informally through the focal points;
   (b) Study should remain the exclusive property of the country so that any release or publication of country data and recommendations of the study should obtain prior formal authorization by the focal points from those countries.

Sri Lanka I
Organizations involved (tentative):

What needed to be done (What was the problem?); Background and justification (*Why* was the problem important?):
   Imbalance between refinery output and the demand for gasoline and middle distillates. There were huge cost implications in either imports or refinery expansion.
How was the work-plan to be implemented?
Step-by-step description and timing:
   (a) Study import of gasoline/diesel price ratios on equipment choices and consumption;
   (b) Identify alternative ratios that were likely to improve the balance;
   (c) Identify alternative measures to bring about changes in demands.

How were results to be monitored and evaluated? (Criteria of success):
[Editor's note: Response not provided.]

Sri Lanka II
Organizations involved (tentative):

What needed to be done (What was the problem?); Background and justification (*Why* was the problem important?):

Pricing of commercial fuels in the rural sector. Kerosene and electrification were substitutes for lighting. Kerosene was subsidized through a ration-coupon scheme. What was (a) the efficiency of the coupon scheme itself; (b) the efficiency of subsidized life-time rates for electricity compared with kerosene?

How was the work-plan to be implemented?
Step-by-step description and timing: How were results to be monitored and evaluated? (Criteria of success):

Study should establish the relative efficiency costs of either subsidy scheme.

Sri Lanka III
Organizations involved (tentative):

What needed to be done (What was the problem?); Background and justification (*Why* was the problem important?):

Efficiency and equity implications of cost recovery in the electricity sector needed to be studied. Very high electricity growth rates had been encountered in the recent past, representing pressures on oil-based thermal generating plus costlier new hydropower. The issue of cost recovery and resource mobilization needed to be researched.

How was the work plan to be implemented?
Step-by-step description and timing:
  (1) Evaluate the revenue levels required to finance operation and investment.
  (2) Compare current tariffs with tariffs needed for efficient price levels.
  (3) Identify equity implications for urban and rural users.
  (4) Identify likely effects of changed prices on demand and conservation.
How were results to be monitored and evaluated? (Criteria of success):
[Editor's note: Response not provided.]

Thailand I
Organizations involved (tentative):
National Energy Administration (NEA).

What needed to be done (what was the problem?); Background and justification (*Why* was the problem important?):
  (a) Do a study of the impact of fuel price changes on the cost of living for consumers of different socio-economic characteristics (income level, family size, location etc.).
  (b) Combine that with the effect of fuel price changes on the choice of fuel in the industrial/agricultural/service sectors.
The recent fuel pricing study by PEIDA did not adequately explore the

socio-economic impact of the proposed fuel price reforms. Since energy prices were distorted and changes appeared to be necessary, it was important to examine the socio-economic impact.

How was the work-plan to be implemented?
Step-by-step description and timing:
A methodology already existed at Cambridge, England, but had been calibrated only for the 1975 input-output table. A 1980 table was now available, and the aim would be to train a Thai researcher in the methodology, and at the same time to update the data, probably at Cambridge. The original consultant would be required to oversee the training and to write the final report.

How were the results to be monitored and evaluated? (Criteria of success):
A test of the success of the project would be that the computer programme could be run in Thailand and the results obtained at Cambridge replicated and that the Thai researcher understood the model and felt confident to update and modify it. The report should be of use to the Thai Government in deciding on its energy pricing policy.

Thailand II
Organizations involved (tentative):
National Energy Administration (NEA).

What needed to be done (What was the problem?); Background and justification (*Why* was the problem important?)
Modify the impact analysis of energy pricing study in Thailand in order to have a full comprehensive energy pricing study.
Recently, Thailand had received a full study of energy pricing by consultants. The study concentrated only on economical and financial points. Thailand did not have a comprehensive study of social impact analysis because of budgetary constraints. Since energy was a very important issue, the Government had to calculate trade-offs among economic, financial and social effects resulting from energy price changes. The Government could use a full comprehensive study as a possible guideline for the implementation of an appropriate energy pricing policy for Thailand.

How was the work plan to be implemented?
Step-by-step description and timing:
The original consultant would be required to modify the social impact analysis along with the available full energy pricing study that Thailand already had.

How were the results to be monitored and evaluated? (Criteria of success):
The report should be of use to the Government of Thailand in its evaluation of its energy pricing policy.

F. Conclusions and recommendations

30. The following general conclusions and recommendations were formulated:

(a) The proposals would be ranked by the ILO/ESCAP team, and in about one month the REDP focal points of the three chosen ones would be contacted to implement them;

(b) More data, budgets, refinements of the proposals were welcomed by the ILO/ESCAP team;

(c) For the other proposals, after collecting more data on budgeting and other refinements, other possible sponsoring agencies would be approached by the REDP secretariat and those studies then might be pursued on a bilateral basis under the general umbrella of the regional energy development programme.

(d) The proposals chosen would have to be considered for official approval by the countries concerned before actual implementation.

G. Adoption of the report (Agenda item 8)

31. The report was adopted on 11 May 1984.

32. A general vote of thanks to the Chairman and all those who made the workshop possible was introduced. The meeting was closed by the Chief of the Natural Resources Division of the ESCAP secretariat.

ANALYSIS OF THE ENERGY PRICING PROBLEM IN CHINA

*Huang Zhi-jie**

In recent years, energy in China has been in short supply and unable to meet the needs of the country's economic development. The way to solve this problem is to exploit and conserve energy with enthusiasm. At present, however, initiatives in the areas of developing certain energy resources, rational utilization of various energies, and implementation of now feasible conservation technology are greatly affected by irrational energy pricing. On the other hand, any readjustment in the price of energy would have impacts on people's living standards and on production costs of all sectors. Therefore, it is necessary to study both energy pricing and its problems.

CURRENT SITUATION OF ENERGY PRICING IN CHINA

Before liberation (1949), the price of coal in China was relatively low due to simple mining equipment and low wages for miners. After liberation, coal mining conditions improved, salaries for miners increased, and production costs became higher, but the price of coal has not been readjusted much and the profit rate in the coal industry has been lower than the average profit rate in other sectors. In the early days of liberation, crude oil output was quite small, only 120,000 tons in 1949. The government, therefore, applied a high pricing policy to limit the consumption of oil products. That is why the price of crude oil was two or three times higher than the international market price, and the prices of oil products were even higher. At present, the price of gasoline price is still higher than the current price

*Deputy Director, Energy Research Institute of the State Economic Commission and Chinese Academy of Science, Beijing, China.

in the world market. In the past decade international oil and natural gas prices increased about ten times, while the price of coal almost doubled. Comparatively, the price of energy in China is one-fourth or one-fifth of the international price. For example, in 1982 China's average coal producer's price was 21.5 yuan renminbi (about US$11) per ton, crude oil 102 yuan renminbi (US$52) per ton, heavy oil 60 yuan renminbi (US$31) per ton, and the international prices were US$55, US$250, and US$220, respectively. The price of diesel oil in China is only 14 to 20 percent that of the international price. Low energy pricing has influenced the import and export of energy and of energy-consuming products.

Another problem with energy pricing in China is the irrational price ratio between different energy resources. For instance, the price ratio among crude oil, gasoline, diesel oil, and heavy oil is 1:1.24:1.28:0.88 in other countries, and 1:5.7:2.4:0.6 in China. Compared with other countries, China's gasoline price is higher, but crude oil and heavy oil prices are lower. The irrational price ratios among various energy sources and between energy and some materials and machinery would cause a series of problems in exploitation and utilization.

IMPACTS OF ENERGY PRICE CHANGES ON ECONOMIC DEVELOPMENT AND ENERGY EXPLOITATION AND UTILIZATION

Readjustments of energy prices play an important role in the development of the energy industry, changes in the energy use structure, promotion of energy conservation, and limitation of certain energy demand. In the 1950s and 1960s, following the rapid development of oil and natural gas industries, the costs of oil and natural gas consumption were lower than coal; thus, oil and gas formed the major part of the energy structure. Since 1973, the two oil crises in the West quickly raised the prices of oil and gas in the world market, which brought many problems to the important oil-consuming countries.

First was the influence of energy on the economic development of these countries. Appendix II. Table 1 shows the average economic growth rates for the seven industrialized countries in different periods. In the 1960s and early 1970s, economic growth rates in these countries were more than twice the average growth rate after the energy crisis. This is because the rising energy price led to an increase in production costs and in product prices, and a decrease in market demand and economic growth rate. At the end of 1982, the oil price per barrel dropped US$5, and economists generally predicted that economic growth rate in the West would increase slightly.

Second was the influence on energy consumption, especially on oil import and consumption. The rapid rise of oil price in the world market made countries that import and consume oil in great amounts control the increase of oil consumption and carry on energy conservation policies, taking oil conservation as the main objective. In recent years, tangible results have been achieved and energy consumption has been reduced. We can see from the data in Appendix II. Table 2 that before 1973 the energy consumption growth rate was high in industrialized countries, except the United Kingdom.

Since 1973, the energy consumption growth rate has decreased several times; some countries even show negative growth rates (see Appendix II. Table 2). Although the economic growth rate has been reduced since 1973 in these countries, elasticities of energy growth relative to economic growth have also been reduced several times. From 1973 to 1980 the average elasticities were less than 0.4.

Third, exploitation and utilization of nuclear and renewable energy have been promoted. Owing to a rapid increase in the prices of oil and gas, the cost of electricity in oil-fired and gas-fired power stations is higher than in nuclear power stations. Because of this, not only industrialized countries but also a number of developing countries have made plans for the development of nuclear power stations. In June 1980, 230 nuclear reactors were operating in 36 countries and regions, and the total capacity was 123 million kilowatts. According to their specific conditions, some countries have made efforts to develop and utilize renewable energy resources. For example, Brazil uses hydroelectricity and domestic alcohol to reduce oil consumption; New Zealand and Mexico have developed geothermal energy; and many industrialized countries use solar energy to provide hot water. Research and development in the utilization of renewable energy (such as wind power, solar energy, ocean energy) has also progressed.

Fourth, the rise in the price of energy has aggravated inflation in most countries. Appendix II. Table 3 shows changes in the average annual growth rate of the consumption cost before and after the oil price increases of 1973 in some countries.

INFLUENCE OF ENERGY PRICE ON ENERGY DEVELOPMENT IN MAJOR INDUSTRIALIZED COUNTRIES

China is a socialist country with a planned economy. Price in China does not play a critical role in supply and demand adjustment as it does in the West. However, the existing energy price influences energy resource development, conservation, and rational utilization. Thus, to accelerate economic development, energy pricing must be done properly.

Low profit rates in the coal industry hold up resource development

Coal is an important energy resource in China, accounting for more than 70 percent in energy structure. The major coal-producing sectors have conducted several price readjustments but have been unable to change the situation of economic loss and low profit rates.

Since 1958, the production cost per ton of coal has increased 13 yuan renminbi, while its price increased only 9 yuan renminbi. The production cost in some coal mines has surpassed the selling price and thus these mines have incurred a loss. Because of this, most of the coal enterprises, although diligently operating their enterprises, are unable to obtain a due profit, and the enthusiasm for production has therefore withered.

Low energy price hinders conservation

In other countries, energy consumption has declined considerably since 1973. One of the most important conservation incentives is a large increase in the price of energy relative to the price of other goods. After the 1973 increase in energy prices, the use of formerly uneconomic savings equipment became economically viable; the utilization of previously too expensive alternative energy also became viable.

Before 1973, the prices of crude oil and oil products in China were much higher than prevailing international price levels. The price of coal was lower, but not by very much. The problem is that compared with the prices of iron and steel, and cement and machinery, and with the price of coal in other countries the price of coal in China is much lower. Thus, some of the conservation technologies, processes, and equipment, which were economically reasonable and feasible abroad before the increase in the price of energy might not have been reasonable, economic, or feasible in China. Now that international energy prices have increased considerably, more conservation technology, processes, and equipment will be economically feasible in China.

For instance, energy prices affect the selection of insulation materials, thickness of insulating layers, and methods of construction. Since 1973, insulating layers in other countries have become thicker, protecting metal enclosures has been popularized in construction, and energy loss from pipelines and equipment has been reduced as much as possible. Because energy prices are low in China, in most cases, we can choose only low-quality insulation materials. Some better quality insulation materials such as rock wool and glass wool, which are widely used abroad, are unmarketable in China because of their relatively higher prices, even though China has the capacity to produce them. The economic insulating layer is thinner in China compared with that which is used abroad, due to the low energy price. In construction, strawrope wrapping or glass wool sheet painting is used instead for protecting metal enclosures, and, according to tests, energy loss doubles with these materials.

Another example is the development of central heating, which is actually a way of exchanging iron and steel for energy. Installation of heating pipeline works would use a lot of steel but could save energy. The price ratio of steel and coal in other countries is several times lower than that in China. Thus, the development of central heating abroad is economically reasonable, but probably uneconomical in China.

In the past, energy was cheap abroad and waste heat recovery was generally not economical. After the energy price increases, not only is use of high-temperature waste heat economical, but in many cases low temperature waste heat recovery with heat pumps is also economically feasible. Because the price of energy is low in China, machinery is relatively expensive, and the economics of some high temperature heat recovery is low, to say nothing of low-temperature waste heat utilization.

Irrational utilization of energy caused by unreasonable price ratios between various kinds of energy

Irrational price ratios between various kinds of energy lead to an irrational and wasteful utilization of energy and at the same time cause environmental pollution. Several examples are given here.

The problems in urban gasification are an example. Both developed and developing countries are moving towards urban gasification. Unreasonable price, however, greatly hinders the development of urban gasification in China. Although LPG for residential use has a considerable result in conservation, its producer price is only 50 yuan renminbi per ton, which is half of the crude oil price and lower than the price of combustion oil. In that case, LPG as the fuel used in refineries would bring greater benefit to enterprises. Before 1973, the price of LPG per ton in the international market was equal to two tons of crude oil. After 1973, although oil prices increased rapidly, the LPG price was generally higher than the crude oil price. Natural gas is both a valuable raw material for the chemical industry and the most perfect fuel for residential use. Countries with rich natural gas resources would use it as raw material for the chemical industry and fuel for residential use, and only the extra amount can be supplied to industry or the power station as fuel. Countries which lack natural gas would import LPG as chemical raw materials and as fuel for urban residences, and its price is almost the same as crude oil at thermal value. The natural gas price in China is less than half that of crude oil and in some places even lower than the price of good quality coal at thermal value. Thus, most of the natural gas used in industries and factories is for fuel. The conservation effects of burning natural gas in these enterprises are quite limited. Natural gas as residential fuel can have a significant conservation effect, improve people's living standards, modernize cities, and reduce pollution to improve the environment as well.

Other examples are gasoline and diesel oil, which are products from crude oil processing. Their production procedures and the amount of energy consumed are basically the same. From the perspective of use value, diesel oil is more efficient, and vehicles using diesel oil can save 30 to 40 percent of oil compared with those using gasoline. At present, however, the price ratio between gasoline and diesel oil is 1:0.4, that is, diesel oil price is more than 50 percent lower than that of gasoline. Refineries are not willing to produce diesel oil because of the low profit it generates. This has led to a reduction in diesel oil and gasoline output ratio in recent years. It was 1.80 in 1979, 1.69 in 1980, and 1.60 in 1981. Unreasonable pricing has hindered the development of diesel oil production and rational utilization of oil products.

Moreover, locomotive power is changing in the direction of the more efficient electric and diesel models in developed countries, as well as in developing countries. In China, steam locomotives have been the most economic for a long time because oil and electricity prices are higher and coal prices lower compared with international prices. An unreasonable energy price ratio has led to an economically rational backward mode of production.

CONCLUSION

1. Since the founding of the new China, energy production and supply have been based on the needs of domestic economic development. Except for exporting a small amount, most of the energy produced in China is to meet domestic needs. In the early days of the new China, required oil products were mainly imported. But because oil consumption represented a very limited share in the energy consumption structure, 95 percent of the energy consumed was provided domestically. In the 1960s, China became self-sufficient in oil with the exploitation of Taching and other oil fields. In the early 1970s, China began to export oil. Before that, China exported a small amount of coal, but more than 95 percent of the energy production was to meet domestic demand. Because of the policy of self-sufficiency, the continuous increase in energy prices in the world since the energy crisis in 1973 has had no influence on the energy price in China.

2. Energy price readjustment helps to accelerate energy conservation. The trend of energy conservation is to improve equipment and use new technology. Investment for coal conservation per ton of coal would increase year by year. If energy price is low, bank loans cannot be paid back with the money saved from energy conservation within the fixed time. For example, the current coal price is 21 yuan renminbi per ton and will be 35 yuan renminbi plus transportation cost. Investment to conserve one ton of coal is 300 yuan renminbi; the annual bank loan interest is 5 percent. Using compounded interest, the project will be completed in two years and capital will be paid back from the third year, after which 15 years will be required to pay back capital and interest. Therefore, enterprises will not be willing to conserve energy. In view of this, energy price readjustment would make

some conservation projects economically favourable; oil-conserving projects would be especially attractive to enterprises.

3. Energy price readjustment is favourable to rational utilization of all kinds of energy. Energy — such as coal, oil, and natural gas — has different usages, effects, and economics. Unreasonable price ratios among energy sources would lead to a waste in utilization. Thus, a readjustment of the ratio between energy prices could lead to a more rational use of energy.

CRITERIA FOR ENERGY PRICING POLICY

Appendix II. Table 1 The average economic growth rates in major industrialized countries

Country	Average GNP growth rate (percentage)		
	1960–70	1970–73	1973–80
United States	3.97	4.72	2.39
United Kingdom	2.79	4.43	0.83
Germany, Federal Republic of	4.85	3.96	2.35
France	5.78	5.56	2.90
Italy	5.66	3.86	2.78
Netherlands	5.13	4.59	2.16
Japan	11.00	8.23	3.79

Appendix II. Table 2 Changes in energy consumption in seven countries

Country	Average growth rate of energy consumption (percentage)		
	1960–70	1970–73	1973–80
United States	4.2	3.26	−0.13
United Kingdom	1.78	1.82	−1.59
Germany, Federal Republic of	5.67	4.58	−0.46
France	6.21	6.58	−0.27
Italy	9.91	5.26	0.61
Netherlands	9.25	9.56	0.73
Japan	13.55	7.02	1.45

Appendix II. Table 3 Average annual growth rate of consumption cost (percentage)

Country	1961–70	1971–73	1974–78
United States	2.6	4.5	7.3
Japan	5.3	7.4	11.2
Germany, Federal Republic of	3.6	5.3	4.5
France	4.0	6.3	10.7
United Kingdom	3.9	8.5	16.0

Appendix III

ENERGY PRICING IN NEPAL

*S. N. Sharma**

INTRODUCTION

By almost any standard Nepal is one of the least developed countries in the world. The per capita annual gross domestic product (GDP) is currently estimated at US$140 in 1980 dollars; there are only six nations in the World Bank's *World Development Report* with a lower per capita GDP. The physical and other obstacles to development are much more severe than those found in most other countries. Nepal is facing rapid population growth, a relatively narrow resource base, the extreme inaccessibility of many parts of the country, a landlocked position, and a relatively inexperienced administration.

The economy of Γ ɔpal is dominated by the agrarian sector which employs more than 90 percent of the economically active people, comprises 70 percent of the value of exports, and represents 65 percent of GDP. Nepal's comparative advantage in agriculture reflects the lack of natural resources that might serve as the base for industry and an untrained human resources base that limits its industrial competitiveness.

The energy consumption pattern in Nepal mirrors the rural agricultural nature of its economy and the small size of its modern industrial and commercial sectors. Currently, the major infrastructure requirements lie in the energy sector. The exploitation of the forests for fuelwood has created the spectre of ecological disaster; large-scale afforestation and reforestation programmes are thus essential, along with ingenuity in satisfying rural energy needs. Developing the abundant indigenous hydropower resource

*Mr. Shiba Nathat Sharma, Executive Director, Water and Energy Commission, Naya Baneswar, Kathmandu, Nepal.

is another formidable task. If the energy constraints to industrial expansion are to be alleviated and the demands of urban residential users satisfied, Nepal needs a large-scale reliable generating capability and a more extensive transmission and distribution network. Thus, the energy problem is twofold. Rural areas depend on traditional fuels for virtually all of their energy requirements. In these areas the growing population and agricultural demands have placed the forest under heavy pressure. The second aspect of the problem facing Nepal is to ensure that the energy demands of the slowly developing modern sector, in particular the demand for electricity, can be supplied.

EXISTING CONSUMPTION PATTERNS

Consumption characteristics

The average annual per capita consumption of commercial energy in Nepal is equivalent to 16.5 kilograms (kg) of coal equivalent. This consumption is concentrated in the urban centres as the topography precludes widespread distribution networks. The 1978 United Nations estimate[1] of commercial consumption for all developing countries is 449 kg of coal equivalent per capita. Nepal's low level of commercial energy consumption reflects of an economy at a low level of economic growth and with a structure which is dominated by an agricultural sector based on traditional farming techniques. The current level of consumption is also inconsistent with the nation's natural energy endowment in the form of hydropower.

The second major characteristic of energy consumption is the total domination of the sector by traditional energy forms, firewood in particular. More than 90 percent of the total consumption is in the form of traditional fuels for domestic use — mainly cooking. The energy sector displays an almost total reliance on its least plentiful resources — wood and foreign exchange for imported oil — and virtually no use of its most abundant energy resources — hydropower and solar.

The modern sector of the Nepalese economy accounts for less than 5 percent of the total energy consumption. This demand is shared equally between transport and industry. The severe topography of the country restricts both industrial expansion and the widespread use of commercial energy to the terai[2] and to the Kathmandu Valley. Despite the fact that some 60 percent of the population live in the hill areas, there is not a sufficiently developed transport network for large-scale goods or fuel movements to the largely remote hill communities.

Traditional fuels

Three traditional fuels — fuelwood, agricultural wastes, and animal dung — provide the vast majority of energy consumed in Nepal. In the absence of adequate fuel substitutes in terms of both quantity and price, these energy forms will continue to dominate the energy sector. It is unlikely that even intensified exploitation of hydro-based electricity would be able to make a significant impact on the short- and medium-term consumption of traditional fuels.

Firewood

Wood has been and will continue to be, for the foreseeable future, the major energy form consumed in Nepal. The reasons for this are threefold:

1. It has been readily available in apparently unrestricted quantities throughout most of the country in close proximity to the points of consumption.
2. It is perceived as a free good with no direct cost except the time and effort to collect it.
3. It requires no major capital investment nor advanced technology in its exploitation and use.

As a result of the above, the use of wood has been unchecked for centuries, but only recently has the growth in consumption outstripped natural regeneration. With continued current rates of growth, studies indicate that it will take only a few decades before the tree cover will have virtually disappeared in many areas of the country and become extremely remote in the remainder.

The vast majority of the fuelwood consumption in Nepal occurs in the rural areas where it is virtually the only fuel available. Less than 1 percent of the total fuelwood consumed is utilized for purposes other than domestic cooking and heating. Surveys indicate that brick kilns dominate the industrial portion of this demand, and that restaurants and sweet shops dominate the commercial portion. It is anticipated that this consumption pattern has not significantly altered. Wood comprises 93 percent of the total energy consumption.

Agricultural and animal wastes

Agricultural waste is a traditional fuel widely used for cooking and heating in Nepal. It includes hay, husks, crop residues, grasses, leaves, sticks, and bark and represents 1.6 percent of the total energy consumption in the country. Animal dung has long been used by the people of Nepal as a fuel for cooking in the form of dried dung-cakes. Animal dung, which contributes 0.6 percent of the national energy balance, is used for fuel almost exclusively in the terai where the cattle population is large and alternative fertilizers are readily available. The use of dung in the hills is predominantly as a fertilizer; however, as forest areas decline, dung-cakes are being used increasingly as a cooking and heating fuel.

Commercial sources of energy

Three commercial fuels are used in Nepal: petroleum-based products, coal, and electricity. They currently represent 5 percent of total energy consumption and are predominately "urban" fuels. Despite their price when compared with that of traditional fuels, which have the appearance of being free, commercial energy consumption can be expected to increase because of their high calorific values, ease of use, and relatively high end-use efficiency. The major drawbacks to their widespread adoption are the problems associated with distribution and the high cost of equipment with which to utilize the energy.

Petroleum products

All petroleum products used in Nepal are imported from India. There are no identified, commercially exploitable oil or natural gas deposits although seepages of hydrocarbons do occur and some survey work is under way. Crude oil is purchased on the open market by the Nepal Oil Corporation for delivery to India; current suppliers are Saudi Arabia, Iraq, and the Soviet Union. In exchange India supplies the required mix of petrol, diesel oil, kerosene, furnace oil, and aviation fuels from refineries near the Nepal border according to an agreed pricing formula. The corporation imported some 139,300 kilolitres of products in 1982–83. This volume and the observed mix of products (i.e., no heavy products) warrants neither a product pipeline from, say Calcutta, nor a crude oil pipeline to a refinery in Nepal. In addition, Nepal Gas Industry imported an estimated 706 tons of LPG in 1982–83.

Nepal's annual per capita oil consumption is some 11.6 kg of coal equivalent, one of the lowest in the world. This statistic is both an indicator of the country's lack of modern infrastructure and its deficiencies in manufacturing and commercial potential and output. The overall rate of growth of petroleum fuels consumption is 7.6 percent per annum and is in line with that of most oil-importing developing countries. It is a reflection of both the low level of consumption and the fact that a country like Nepal, finding itself in the initial stages of development, is unlikely to be able to significantly reduce its dependence on oil without seriously retarding its capability to develop. Petroleum products represent 3.6 percent of the overall energy balance. Sixty percent of petroleum consumption is for transport and 26 percent for domestic use.

Virtually all of the coal consumed in Nepal is imported from India. Peat and lignite occurrences in the Kathmandu Valley are not viable for large-scale exploitation but do provide approximately 5,500 tons annually for local use, mainly to the brick industry.

Imports of coal into Nepal fall under two broad categories: (1) imports under the quota system, and (2) imports obtained on the Indian free market. In the first case, prices paid for coal are somewhat below world prices; in the latter case, world market prices apply. There is very little coal imported outside of the quota system.

Supplies under the quota system are both erratic and unreliable, and quotas often go unfilled. The supply of coal is affected by its availability in India, labour, technical problems linked with the Indian railway system, and the readiness of the authorities to sell such fuel to Nepal at concessional prices. The quota for 1980 was 112,000 tons and included steam and slack coal, hard and soft cokes, and low-grade coke. It is estimated that some 37,000 tons of coal were imported, representing approximately 1 percent of total energy. Coal in Nepal is consumed for three major purposes: transport (13 percent), manufacturing (84 percent), and industrial power generation (3 percent).

Electricity

Electricity is a particularly important energy form in Nepal. Hydropower is the country's only known source of commercial energy, and the potential for development is considerable. Currently the consumption of electricity is restricted to major urban areas where population density and nearby generating sites have justified the investment. The level of consumption is expanding rapidly from this small base. The high cost of distribution is the major impediment to the widespread electrification of the rural hill areas.

A total of 163.9 gigawatt hours (GWh) of electricity were sold in 1980–81 within Nepal, less than 0.5 percent of total energy consumption. The overall rate of growth for electrical energy is in excess of 15 percent per year, reflecting the small consumption base and a country at the initial stages of development.

In Nepal, electricity is supplied by public and private utilities, by private companies that produce power for their own use and also through imports from India.

ENERGY PRICING, RESOURCE COSTS, AND PRICING POLICY

The subsistence nature of Nepal's economy, and the severe topographical restrictions placed on rural fuel distribution networks, result in more than 90 percent of the energy consumption in Nepal not being traded within the monetized sector of the economy. Rural energy, which is primarily fuelwood, is not supplied through a market system but rather is collected by the user on a day-to-day basis for immediate consumption. Urban consumers typically have a wider variety of energy forms at their disposal and have an established market system through which to purchase them.

This pricing structure, by definition, lends itself to overall pricing policies which neither reflect long-run marginal costs nor promote use of the most abundant resources. Two price structures, rural and urban, must be addressed.

Fuelwood

Market prices

Urban fuelwood for both domestic and industrial use is supplied by the Fuelwood Corporation (FCN), by private contractors, and by the Department of Forests. Commercial extraction of fuelwood from the forests is subject to a royalty payment of NRs 40 per ton. This cost, however, represents less than 10 percent of FCN cost of supply to the Kathmandu Valley. Transport charges represent fully 70 percent of the total cost. In 1981, the FCN price for fuelwood in the valley ranged from NRs 420 to 530 per ton, depending on location. These prices are only slightly above costs. Prices in the terai were approximately half of this figure reflecting the substantially lower transport component. FCN pricing policy is to set prices to cover production costs. These costs, however, do not include a value for the wood itself; therefore, the prices provide an implicit subsidy to urban fuelwood consumers at the expense of long-term forest stability.

The increasing difficulties in obtaining fuelwood have increased the price for fuelwood supplied by private contractors to NRs 720–870 per ton in the Kathmandu Valley (NRs 330–400 per ton in the terai). It is understood that His Majesty's Government has decided to restrict or eliminate forest access to private contractors. If this happens, FCN would effectively be the only bulk supplier for the urban areas. It is not known how this will affect fuelwood market prices.

FCN supplied approximately 25 percent of the fuelwood demand in the six major urban centres in 1981. This represented 43 thousand tons of wood. An additional 96 thousand tons were supplied to other urban centres in the terai and 144 thousand tons exported to India.

As previously noted, there is no established market system for rural fuelwood consumption. The "market" price is the perceived value of the labour required to collect the daily fuel requirement and carry it to the household. Much of this labour is supplied by children too young to work in the fields. Within such a system it is impossible for the government to influence fuelwood consumption levels or promote interfuel substitution (if alternatives were available) through a pricing system. Fuelwood conservation through improved stoves, and forestry programmes to provide long-term supply will, in the absence of market prices, only be successful if their installation and implementation costs are less than the perceived cost of collection. The rural fuelwood subsidy is again at the expense of long-term forest stability.

Economic resource cost of fuelwood
The economic resource cost of fuelwood is determined as the least cost means of providing a ready supply of wood for consumption. This figure determines the value to the economy of not utilizing wood in that it represents the cost of supplying an extra cubic metre of fuelwood.

Long-term cost. The long-run resource cost is based on the costs and benefits associated with the Government of Nepal/UNDP/FAO Community Forestry Development Project. The major afforestation/reforestation components of the project expect to develop 51,750 hectares of productive forests over the 5-year term of the project. Three classifications of forest are included: Panchayat forests, Panchayat protected forests, and private plantings. Each classification indicates a different level of responsibility and product distribution, as well as differing yields and forest management techniques. Based on an annual discount rate of 12 percent, the economic resource cost is calculated at NRs 423 per cubic metre (NRs 0.6/kg). This analysis slightly overestimates the value of fuelwood for it does not include the grass and leaf fodder in the production stream, while the costs associated with fodder production are implicitly included in the cost figures. It is, therefore, a conservative estimate when used in a comparative analysis. Clearly, reforestation is a long-term supply option with little or no production during the early years. Therefore, it would be inappropriate to use this figure in the analysis of energy supply options for the immediate future.

Short-term cost. The short-term resource cost of fuelwood is based on the value of dung as a substitute for fuelwood. The value of dung, which also does not have a clearly defined market price, is in turn determined on the basis of avoided losses to maize production. The analysis is based on the assumption that the labour involved in collecting and preparing animal dung for use as a fuel is the same as that required to compost and spread an equal amount of dung on the fields for crop fertilization. The second major assumption is that Nepal will continue to be a net importer of foodgrains over the short run and that a loss in production will be offset by increased imports. Two resource costs have been determined. The first is for the remote hill areas and includes a large transportation component, the second is applicable to the readily accessible terai and low hill areas and excludes much of the transport costs. The short-term resource costs are NRs 488 per cubic metre for the terai and low hills areas and NRs 885 per cubic metre in the remote hills.

An alternative method of determining the economic resource cost is based on the cost of importing chemical fertilizers of equivalent nutrient value to replace the dung which has been diverted from the fields. The basic assumption underlying this computation is that the fertilizer would indeed be imported. The short-term resource costs using this method are NRs 593 per cubic metre for the terai and low hills and NRs 1,004 per cubic metre in the remote hills.

Chemical fertilizers are not in widespread use in Nepal. There is no doubt that, with correct use, chemical fertilizers can significantly increase crop yields and greatly reduce the need for compost. However, unlike compost, fertilizers allow little margin for error. Errors in application timing, amount, or poor water management can eliminate or reverse anticipated production response. In addition, the introduction of a cash import to a mainly subsistence farming system may introduce risks which a farmer is unwilling to take. It would also introduce an element of foreign input dependence that the nation may be unwilling to accept. Once fertilizer use has started, compost management is largely eliminated and difficult to reestablish in the event of fertilizer supply constraints.

The introduction of a chemical fertilizer-based farming system would require that the agricultural administration be changed to establish a sufficient infrastructure to accommodate the more advanced system. For these reasons it is unlikely that the basic assumption in the resource cost calculation with regard to fertilizer imports will be fulfilled. The resource costs derived from maize imports are therefore used for comparative purposes.

Petroleum products

All petroleum products consumed in Nepal are imported by the Nepal Oil Corporation and distributed by its licensed dealers. The current import arrangements are such that Nepal purchases crude oil and products and delivers them to an Indian port. In exchange, India provides an equivalent amount of refined products from refineries located close to the Nepalese border. The retail prices of petroleum products are reflective of international prices, and direct subsidies through the general pricing system are not in evidence. In fact, taxes and duties on petroleum products, in particular motor spirit, provide a substantial source of government revenue. Motor spirit is subject to taxes and duties equivalent to 112 percent of pretax costs; diesel oil is taxed at 30 percent. Kerosene is subject to only a 12 percent tax, in part in order to "subsidize" the energy costs of low-income, predominately urban, families.

Coal

Coal imports to Nepal have two sets of prices. Imports under the quota system are charged at concessionary prices (NRs 360/ton), while imports outside of the quota system are at much higher international prices (up to NRs 2,000/ton). There are no customs duties or import taxes on coal imports.

Electricity

Market prices

The Nepal Electricity Corporation (NEC) operates the bulk of the nation's generating and distribution facilities. Virtually all of the consumption of electricity is in the urban centres which restricts access to less than 7 percent of the total population. The country's topography precludes a widespread distribution network beyond that which could be developed in the terai. This means that approximately half of the people of Nepal can expect no ready access to electricity.

The NEC tariff history is a reflection of the type of expansion on the system. Most of the new plants were hydraulic, which reduced annual operating expenditures, while the bulk of the expansion capital was made available on a grant basis. The resulting tariffs have always been significantly below the marginal costs of power production. In 1971, the NEC reduced its electricity tariffs by an average of 43 percent to provide a reasonable cost alternative to kerosene which was in short supply. This reduction was never rescinded with the result that electricity prices represent approximately a 50 percent subsidy to urban energy consumers. In addition, the high level of system losses further erodes the corporation's financial base.

A tariff increase of 58 percent in April 1983 and a proposed further increase of 65 percent will bring revenues in line with a covenant included in recent financing arrangements which require a 6 percent rate of return on NEC's operations. These tariffs are still well below the long-run marginal costs of providing electricity in Nepal.

Economic costs

Major generation. The marginal costs of providing electrical power by season and time of day have been determined for the Nepal Power System. The overall philosophy of the marginal cost approach is that the consumer should be charged at a rate which reflects the true economic costs of supplying power to that consumer. With this information the consumer can decide whether consumption at a particular time is worth the charge he is facing. The result should be the most efficient allocation of resources.

The costs determined in the tariff study include an allowance for the location of the load (urban or rural), the size of the load (11 kilovolts (kV), 33 kV per 400/230 V), the season (dry or wet), and the time of day (peak or off-peak). The average marginal costs associated with grid supply are considerably higher than current tariffs.

Micro-hydro generation. The basic capacity cost for a mini-hydro station has been taken at US$2,500 to US$3,500 per installed kilowatt. The lower figure has been assumed for the terai and lower hill installation, the higher for remote hill sites. A load factor of 20 percent and an economic life of 26 years have also been assumed. At an average cost of 12 percent capital, the unit costs of electricity are NRs 2.6/kWh for the terai and NRs 3.6/kWh for the remote hill areas, respectively.

Ability to pay for energy

There are large components of implicit and explicit subsidy included in the pricing structures for fuelwood and electricity. From the viewpoint of economic efficiency, it would be desirable to increase prices to the level of the long-run marginal costs to enable consumers to determine the value they place on additional consumption with respect to the cost (to the nation) of providing the additional energy. In the final price structure, however, other considerations must be included in the analysis, one of which is the ability to pay for energy. If energy is to be considered as a basic amenity to be provided to a majority of the population, it is necessary that it be priced so that people can afford to buy it.

Most of this analysis is directed at the urban energy sector which is based to a great degree on a market system with viable energy alternatives. The rural sector has few, if any, alternatives to the consumption of wood and includes in its prices a measure of the effort required to obtain the household's energy requirements. If it were law or common practice that two trees be planted for every tree harvested, the rural pricing structure would more accurately reflect the economic costs of fuelwood consumption as it would then include a component of resource cost.

Urban household expenditures for fuel, light, and water amount to 5.6 percent of total household expenditures. This average ranges from 10 percent for the lowest expenditure level to 4 percent for the highest level. Expenditures for electricity show a somewhat different pattern in that the lowest and the highest expenditure levels spend the smallest proportion.

As a percentage of total energy expenditures, expenditure on electricity is estimated to vary between 20 to 55 percent. Electricity expenditures are highest as a percentage of total energy expenditures for the households in the highest expenditure groups, which suggests a greater substitution of electricity for alternative energies among higher income groups. An analysis of electricity consumption reveals that 45 percent of domestic consumers use less than 25 kWh per month (30 percent use less than 15 kWh per month) and only 16 percent exceed 100 kWh per month. For a large percentage of households, the low level of consumption indicates that the primary use of electricity is for lighting.

As a result of these consumption and expenditure patterns, it can be seen that any general increases in the prices of fuelwood and electricity to eliminate the subsidy aspects would amount to an income transfer from the poor to the rich. In addition one would expect a large proportion of low-income consumers to be excluded from further use of electricity and severely restricted in fuelwood use.

Fuelwood price increases must be accompanied by widespread conservation programmes to decrease or at worst maintain the current level of expenditure. Electricity tariff increases should be implemented so as to retain the "lifeline" tariff for low-income lighting and to adjust tariffs for the high levels of consumption to the long-run marginal cost. There is a cross-subsidy between high consumption consumers and low consumers. The top 15 percent of consumers account for 50 percent of total expenditures, while the bottom 25 percent account for only 5 percent. Ignoring the impact of changes in price on consumption, a 50 percent subsidy to the bottom 25 percent of households could be recovered by a 5 percent increase to the top 15 percent of households.

Interfuel price comparison

Comparisons of fuel price depend not only on the absolute price of the fuel but also on availability, the end-use efficiency, and the location of uses. In this analysis we have considered the two most common energy uses — cooking and lighting — and three locations — the terai (including the low hills), the remote hills, and Kathmandu.

Cooking

Ninety-five percent of all energy use in Nepal is domestic, and cooking represents virtually the total demand. Lighting and direct heating demands are negligible. Ninety-six percent of the domestic use is supplied by fuelwood.

Several substitutes for fuelwood cooking are feasible, namely, electricity, LPG, kerosene, and biogas. All have technical problems which affect their application in the remote hills. In particular, kerosene and LPG are subject to high transportation charges; electricity is subject to the same charges during construction for micro-hydro, or to high transmission costs if centralized generators are used; and biogas plants have a decreasing production capability at higher altitudes (and lower temperatures). The economic resource costs of the substitutes for fuelwood cooking are shown in Appendix III. Table 1. The costs are based on the preceding analysis. Kerosene and liquefied petroleum gases have been valued at NRs 4.4 per litre and NRs 8.8 per kilogram, respectively. These are the estimated border prices and are used as the terai prices. The Kathmandu economic costs include a transport charge of NRs 270 per ton. The values for the remote hills are based on the above terai and lower hills price, plus a transport differential equal to six days porterage or NRs 3,335 per ton. This is converted to NRs 2.60 per litre for kerosene and NRs 3.4 per kg for LPG.

Based on the economic resource cost of fuelwood through reforestation, the only long-term competitive substitutes are biogas and improved stoves. The short-run substitutes are these two, plus LPG. Electricity costs would have to be reduced by 60 to 70 percent before this energy form could be considered competitive. Kerosene is competitive only in Kathmandu.

The basic conclusion is that the long-term "solution" to the forestry crisis in Nepal is reforestation as it represents the only feasible method of assuring long-term supply. The short-term crisis cannot be solved merely by introducing substitutes. While this approach may be warranted in isolated cases, the widespread use of substitutes is generally not technically or politically feasible. Biogas plants, for example, are economically attractive, yet their use is technically restricted to less than half of the population because of altitude and only to those households with sufficient cattle (four to six per unit) to operate the system. An additional consideration, however, is that the smoke from the existing stoves plays an important role in the control of insect infestation, both in the roof and in the crops suspended from the roof for drying. Widespread biogas use, particularly in the terai, would require alternative control measures.

The use of LPG as a substitute is economically feasible based on this analysis; however, increased dependence upon a foreign source of fuel does little to promote the government policy of self-sufficiency in energy[2] and would further aggravate the current trade deficit. As in the case of biogas, the introduction of LPG stoves would also involve alternative insect control measures.

Fuelwood conservation in the form of improved stoves and fuelwood drying offers almost immediate relief to the fuelwood crisis. In laboratory research the improved stoves have displayed an overall efficiency of 30 percent compared with 15 percent in a traditional stove. This is a savings of 50 percent. If in practical application this savings is of the order of 30 percent, it could mean a reduction of 1.1 cubic metres of fuelwood per household per year. This is valued at NRs 490 using the economic resource cost from reforestation. The capital cost of the improved stove is approximately NRs 70 plus transportation. Whatever the stove design, the proper drying of fuelwood to a moisture content of 20 to 25 percent will reduce the quantity of wood needed to a given heating requirement by some 15 percent. Reducing the moisture content is desirable for two reasons: to reduce handling and transport costs, and to increase its fuel value. For this report the calorific value of fuelwood is 4,000 kcal/kg, which corresponds to a moisture content of 15 percent on a dry-wood basis. Green wood with, say, a 100 percent moisture content has a calorific value of 2,000 kcal/kg. The difference represents the energy required to vapourize the water and also reflects a change in weight. The costs associated with this conservation technique involve a proper storage facility for approximately a 6-month supply of fuel and the value of that supply.

Lighting

The other major domestic use of energy under consideration is lighting. There are two sources of lighting energy: electricity and kerosene. Neither is readily available in other than major urban centres. In other areas minimal lighting levels are provided by candles, reflected light from cooking fires, and, on occasion, rudimentary kerosene lamps. Annex Table 2 outlines the energy requirements and the costs associated with lighting.

For lighting, electricity, even at the long-run marginal cost of production is the least-cost fuel, while kerosene is the most expensive energy form.

CONCLUSIONS

The following observations are relevant to the analysis of energy pricing in Nepal:

1. More than 90 percent of the energy consumed in Nepal is not traded through an established market system but rather is collected by the user for immediate consumption. It is unlikely that any government action to impose taxes or a pricing mechanism on this energy would meet with any success. First, the costs of controlling such a system would exceed any potential revenue, and second, the subsistence nature of the rural economy precludes any significant level of monetized transactions. A programme to install improved stoves, free of charge, and to train villagers in basic forest management would have far better returns. A requirement to plant two trees for each one cut would help to instill the concept of the resource

cost of fuelwood consumption.

2. Urban energy consumption does depend on a market system for the distribution of energy products. These products, however, are generally traded at subsidized prices, petroleum being the single exception. Fuelwood prices include only the costs of extraction and transportation and exclude resource costs. Coal is purchased at concessionary prices and is often in short supply because of this. Electricity tariffs are well below both those required for a reasonable rate of return on assets and those based on long-run marginal costs.

3. Although fuelwood price increases to incorporate the resource cost of fuelwood would promote economic efficiency, it is also likely that they would impose undue hardship on low-income consumers who financially have little choice among energy sources. An improved stove programme in concert with any price increase would be required to maintain stability in this sector.

4. Electricity tariffs which currently benefit the more affluent consumers must be adjusted to reflect the long-run marginal cost of providing power. Electricity is the most economically efficient fuel for lighting but not for cooking; therefore, consumers who are willing (or only able) to purchase the minimum electricity requirements to satisfy lighting needs should receive this electricity at subsidized rates. This would also promote electric over kerosene lighting. Large household users of electricity who are cooking with electricity should be charged the full marginal cost for this consumption. In doing this, however, a mechanism must be developed to stop the substitution of fuelwood (if still sold at subsidized prices) for electricity in cooking.

5. Biogas plants must be promoted through pricing (by, for example, eliminating taxes and duty on imported components), particularly in less affluent urban areas in order to present a viable alternative to electricity (in terms of price) and fuelwood (in terms of efficiency) for cooking.

Appendix III. Table 1 Resource cost of fuelwood cooking using various substitutes[a]

| | Fuelwood supply | Substitute energy form | | | | | Biogas[d] | |
		Isolated	Grid	Kerosene	LPG	Improved Stove[c]	Family	Community
Calorific value (kcal)	4000/kg	860/kWh	860/kWh	8660/1	11760/kg	4000/kg	5650/m³	5650/m³
End-use efficiency (%)	15	80	80	30	70	21	60	60
Equivalence to one cubic metre of fuelwood[b]	700	610	610	162 1	51 kg	500 kg	124 m³	124 m³
Capacity cost: Terai	—	35500/kW	—	—	—	—	10418	38587
Hills	—	49700/kW	—	—	—	—	—	—
Economic life (years)	—	25	—	—	—	—	15	15
Annual cost @ 12%: Terai	—	4525/kW	—	—	—	—	1565	5790
Hills	—	6335/kW	—	—	—	—	—	—
Load factor (%)	—	20	—	—	—	—	—	—
Annual production	—	1750/kWh	—	—	—	—	869m³	3619m³
Unit cost: Terai	—	2.6/kWh	2.6/kWh	4.4/1	8.7/kg	—	1.8m³	1.6m³
Hills	—	3.6/kWh	—	7.0/1	12.1/kg	—	—	—
Kathmandu	—	—	2.4/kWh	4.6/1	9.0/kg	—	—	—
Economic cost of one cubic metre of fuelwood equivalence								
Terai	488	1578	1586	713	445	346	223	198
Hills	885	2209	—	1134	620	628	—	—
Kathmandu	740	—	1464	745	460	525	—	—

Source: Ministry of Water Resources (1983).

a All costs in Nepalese rupees.
b 420,000 kcal per cubic metre in cooking.
c Assumes a 30 percent saving.
d Additional capital expenditures are required in years 8 and 12.

Appendix III. Table 2 Energy requirements and costs associated with lighting

Energy form	Required energy[a] (kcal)	Economic cost (NRs/hr)		
		Terai	Hills	Kathmandu
Electricity	86	.26	—	.24
Kerosene	1,083	.55	.88	.57
LPG	549	.41	.57	.42
Biogas	1,100	.33	—	—

[a] 100W equivalence (167 candela), based on kerosene pressure lamps at 125 ml per hour for 167 candela, LPG pressure lamps at 14 oz per hour for 167 candela, and biogas mantle at 0.08 m^3 per hour for 66.8 candela.

Source: Ministry of Water Resources (1984).

NOTES

[1]See United Nations (1979).

[2]The term ''terai'' refers to the rural flatlands and low hills lying between the urban area and the remote, higher hills.

[3]The analysis does not include an explicit cost for new appliances. Burners for biogas and kerosene, for example, can be purchased for 75 to 100 Nepalese rupees each.

REFERENCES

Ministry of Water Resources (1983). *Energy Sector Synopsis* Report (June).
Ministry of Water Resources (1984). *Biogas Resources of Nepal* (April).
United Nations (1979). *World Energy Supplies,* 1973–1978, Series J. No. 22.
World Bank (1983). World Development Report.

ENERGY PRICING POLICY WORKSHOP
Bangkok, Thailand, 8–11 May 1984

LIST OF PARTICIPANTS

Members

CHINA

Huang Zhijie
Deputy Director, Energy Research Institute
State Economic Commission
Beijing, China

INDIA

A. K. Mago
Director, Department of Power
Ministry of Energy
Delhi, India

INDONESIA

Munawar Amarullah
Senior Staff for PLN (State Electricity
Company) and National Planning Board
Jakarta, Indonesia

NEPAL

Shiva Nath Sharma
Executive Director
Water and Energy Commission
Kathmandu, Nepal

PHILIPPINES

Gary S. Makasiar
Division for Planning, Ministry of Energy
Manila, Philippines

SRI LANKA

N. Mathanagopalan
Commercial Engineer (Tariffs and Marketing)
Commercial Division
Ceylon Electricity Board
Colombo, Sri Lanka

THAILAND

Sawanee Saratunti
National Energy Administration
Bangkok, Thailand

Siripron Sophidpakdeepong
Senior Economist
Chief of the Electricity Policy
Policy and Planning Division
National Energy Administration
Bangkok, Thailand

Consultants

Corazón M. Siddayao
Research Coordinator, Energy and
Industrialization Project, Resource Systems
Institute, East-West Center
Affiliate Professor, Department of Economics,
University of Hawaii
Honolulu, Hawaii 96848

Gunter Schramm
Advisor
The World Bank
Washington, D.C. 20433

Mohan Munasinghe
Senior Energy Advisor to the President of Sri
 Lanka
Colombo, Sri Lanka

Ramesh Bhatia
Professor of Economics
Institute of Economic Growth
Delhi, India

David M. G. Newbery
Lecturer in Economics
Cambridge University
Churchill College
Cambridge, England

Observer

Donald Hertzmark
Asian Institute of Technology (AIT)
Bangkok, Thailand

K. Goldsmith
Adviser, NRED/DTCD
Department of Technical Co-operation for
 Development
United Nations
New York, New York

Specialized Agency

Eddy Lee
Chief
International Labour Organisation
Asian Regional Team for Employment
 Promotion
Bangkok, Thailand

Rashid Amjad
Senior Development Economist
International Labour Organisation
ARTEP
Bangkok, Thailand

M. S. Kumar (University of Cambridge)
Consultant
International Labour Organisation
ARTEP
Bangkok, Thailand

Maurice DeWuff
International Labour Organisation
Bangkok, Thailand

Intergovernmental organizations

Florian Sauter-Servaes
Assistant Professor
Asian Institute of Technology
Bangkok, Thailand

Ashok V. Desai
Coordinator, Energy Research Group
International Development Research Centre
 (IDRC)
Ottawa, Canada

ABOUT THE AUTHORS

Ramesh Bhatia
Dr. Bhatia is Professor of Economics, Institute of Economic Growth, Delhi, India. He has held visiting appointments at Harvard University and other major universities. He has also served as consultant for international organizations, including the World Bank and the United Nations. He has several publications, including recent ones on energy in the developing countries.

Manmohan S. Kumar
Dr. Kumar is a Fellow of Sidney Sussex College, Cambridge, and Research Officer, Department of Applied Economics, University of Cambridge. He is working in the fields of energy economics and industrial economics. He has acted as consultant to the World Bank, the International Labour Organisation, and other international organizations. He has published in his area of specialization, and was an invited author at the Seventh Congress of the International Economic Association. Dr. Kumar wrote the chapter in this volume as a consultant for the International Labour Organisation/Asian Regional Team for Employment Promotion, Bangkok.

Mohan Munasinghe
Dr. Munasinghe is Senior Energy Advisor to the President of Sri Lanka, and also Senior Economist-Engineer, on leave from the Energy Department of the World Bank. He has taught at universities in Colombo, Sri Lanka, and the United States, and has field experience in energy in over 50 developing countries. He has published extensively in the field of energy economics; among recent publications are: *Electric Pricing*

(Johns Hopkins, 1982) co-authored with Jeremy Warford; *Energy Economics, Demand Management and Conservation Policy* (Van Nostrand Reinhold, 1983) co-authored with Gunter Schramm; and *Costing and Pricing Electricity in Developing Countries* (Asian Development Bank, 1984), co-edited with Shyam Rungta.

David M. Newbery

Dr. Newbery is a Fellow of Churchill College, Cambridge, and Lecturer in Economics, University of Cambridge. He held visiting appointments at Yale, Stanford, and Princeton universities, and has co-authored a study for the Electric Power Research Institute entitled *Effects of Risk on Prices and Quantities of Energy Supplies* (EPRI Report EA-700, Vols 1-4, 1978). He has consulted on energy policy for major oil companies, international organisations, and the Royal Institute of International Affairs. He was recently Chief, Public Economics Division at the World Bank, Washington, D.C. He has published extensively on energy, commodities, futures markets, and on tax and price policy, particularly in the energy field.

Corazon M. Siddayao

Dr. Siddayao is currently Research Coordinator, Energy and Industrialization Project, The East-West Center, Hawaii, and Affiliate Professor, Department of Economics, University of Hawaii. She has served as Expert at the U. S. Federal Energy Administration (predecessor to the U. S. Department of Energy) and consultant for the Ford Foundation, the World Bank, the Asian Development Bank, and the United Nations. She also earlier spent ten years in the petroleum industry. She has published extensively in the energy field, including two books on petroleum published by the Oxford University Press (*The Supply of Petroleum Reserves in South-East Asia* and *The Off-shore Petroleum Resources of South-East Asia*) and several articles on both petroleum resources and energy-related external debt and pricing issues.

Gunter Schramm

Dr. Schramm is a senior energy economist at the World Bank, Washington, D.C. Until 1983 he was a professor of resource economics at the University of Michigan, Ann Arbor. He has been a frequent consultant to the World Bank, the Organization of American States, the Asian Development Bank, the United Nations Development Program, AID and many governments throughout the world. He is the author of *The Role of Low-Cost Power in Economic Development* (Arno, 1979), *Asian Energy Problems* (with T.L. Sankar, Praeger, 1982), *Energy Economics, Demand Management and Conservation Policy* (with M. Munasinghe, Van Nostrand Reinhold, 1983) and *Regional Poverty and Change* (ed., Queen's Printer, 1976), as well as of many professional articles and monographs.

INDEX